Quantum Key Distribution Explained

A Beginner's Guide to the Physics of Unbreakable Keys

No physics or math background required

Concept-first throughout

Cybersoft Publishing LLC

Claude J. Louis-Charles, PhD

Quantum Key Distribution Explained: A Beginner's Guide to the Physics of Unbreakable Keys

This book is intended for general educational and professional purposes only. It is not legal, regulatory, or formal cryptographic engineering advice and is not a substitute for guidance from a qualified cryptographer, security architect, or licensed information assurance professional. Specific networks, satellites, and vendors mentioned are illustrative; the author is not affiliated with, endorsed by, or compensated by any vendor named in this book.

First Edition

Published 2026

Contents

Introduction

Every few decades, something changes about how the world keeps secrets. Before the internet, secrets traveled in sealed envelopes and inside locked briefcases. When email arrived, new methods had to be invented quickly so that private messages could cross open networks without being read by strangers. Now something is changing again, and it is bigger than email. A new kind of computer, one that works on the strange rules of the quantum world rather than ordinary electronics, is on the horizon. When it arrives, the math-based locks that protect most of today's digital life will need to be replaced.

This book is about one of the most elegant answers to that challenge: quantum key distribution (QKD). The phrase sounds technical, but the core idea is beautifully simple. Instead of relying on a hard math problem to keep a secret key secure, QKD uses a law of physics: observing something quantum always disturbs it. That one fact, strange and counterintuitive as it may sound, turns out to be enough to build a communication channel in which any eavesdropper is automatically detected and expelled.

You do not need a physics degree to follow this story. You do not need a math background. You do not need any prior experience with security or cryptography. What you need is curiosity and a willingness to sit with ideas that are, at first, a little weird. This book will do the rest.

Who This Book Is For

This book was written for three kinds of people, and all three start from the same place: no background in quantum physics, no background in cryptography, and a genuine desire to understand what the fuss is about.

The first group is managers and decision-makers. If you lead a team, run an organization, or serve on a committee tasked with evaluating quantum security investments, you need more than a vendor pitch. You need to understand what the technology actually does, what it cannot do, and what questions to ask before you sign anything. This book gives you that foundation. By the last chapter, you will be able to hold a real conversation with technical staff, evaluate claims with appropriate skepticism, and make decisions grounded in evidence rather than hype.

The second group is students. Whether you are finishing high school, starting college, or somewhere in the middle of a degree that has little to do with physics or computer science, the quantum revolution is going to shape the world you work in. Understanding QKD at a conceptual level is a genuine career asset. It signals that you can think clearly about unfamiliar ideas, which is exactly the skill that every industry is hungry for right now.

The third group is adult enthusiasts: curious people of any age who read broadly, follow technology news, and want to actually understand what they are reading about rather than skim the headlines. If you have picked up books about artificial intelligence, climate science, or genetics because the subject fascinated you, this book is written in the same spirit. Serious ideas, explained honestly, without talking down to you.

All three groups share one thing. They deserve an explanation that respects their intelligence without assuming their expertise. That is the promise of every page that follows.

Diagram I.1: The three readers this book serves

What You Will Learn

The twelve chapters of this book build on one another the way a staircase builds: each step lifts you a little higher, and none of them ask you to jump. By the time you reach the final chapter, you will have covered a lot of ground. Here is a preview of the journey.

You will start with the basics of encryption itself. Before quantum anything can make sense, you need a clear picture of what a secret key is, why sharing one securely is so hard, and why the solution the internet currently uses carries a hidden risk. The first two chapters do exactly that, and they do it with everyday stories rather than equations.

From there, you will learn just enough quantum physics to understand QKD. Not everything in quantum mechanics, just the handful of ideas that matter for this specific technology: why particles of light are different from billiard balls, what it means for something to be in two states at once, and why the act of measuring a quantum object inevitably changes it. Chapter 3 takes those concepts slowly, with plenty of analogies, and leaves you with a toolkit that is exactly the right size for what comes next.

Chapters 4 and 5 are the heart of the book. Chapter 4 walks you through the most famous QKD protocol (called BB84, named after the year and the paper that introduced it) step by step, in plain language, using a coin-flipping story that makes the logic feel almost obvious in hindsight. Chapter 5 explains the single most remarkable property of the whole system: any eavesdropper who touches the quantum channel leaves a trace, and the two communicating parties can detect that trace and throw away the compromised data. That detection guarantee is what makes QKD different from every encryption method that came before it.

The next section of the book moves from theory to practice. Chapter 6 looks at what QKD hardware actually consists of, from the lasers and fiber-optic cables to the single-photon detectors that make it all work. Chapter 7 tours the ground-based quantum networks that already exist and carry real traffic in several countries. Chapter 8 looks upward, to the satellites that solve one of QKD's biggest technical problems: the fact that optical fiber gets noisy over long distances.

Chapter 9 shifts from "how it works" to "who is using it and why." Banks, government agencies, healthcare networks, and data centers are all named as real user categories, with honest accounts of what problems QKD actually solves for them right now.

Chapter 10 is one of the most important in the book. It covers the honest limitations of QKD: what it cannot do, the remaining attack surfaces, and the practical challenges of deploying it in the real world. No technology deserves a free pass on its weaknesses, and this chapter does not give QKD one.

Chapter 11 introduces the other main answer to the quantum threat: post-quantum cryptography (PQC). QKD and PQC are very different approaches, and this chapter compares them fairly, explains why security experts often advocate using both together, and gives you the vocabulary to follow the ongoing policy debate.

Chapter 12 looks forward. Quantum repeaters, quantum networks that span continents without trusted-node compromises, and the long-horizon possibility of a global quantum internet are all examined honestly, with a clear distinction drawn between what is real today, what is near-term, and what remains a scientific aspiration.

After the twelve chapters, four appendices give you additional resources. The Glossary covers every technical term used in the book with plain-language definitions. Appendix B is a slow-motion, pencil-and-paper walkthrough of BB84 for readers who want to see the math behind the magic. Appendix C is a curated guide to free online simulators and videos. Appendix D is a staged reading path for anyone who wants to go deeper after finishing this book.

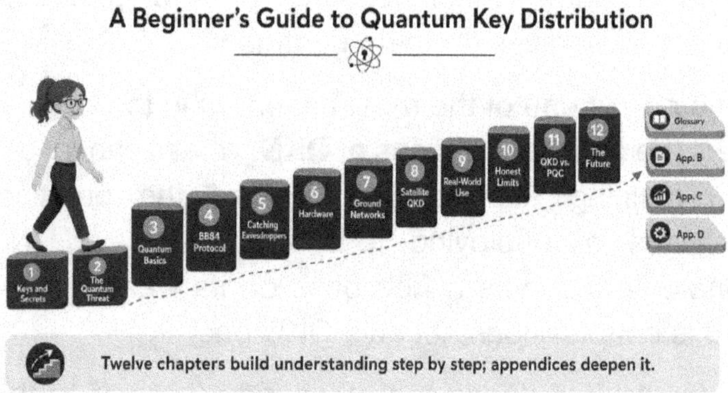

Diagram I.2: The book roadmap: twelve chapters and beyond

How to Read This Book

The simplest way to read this book is from beginning to end. Each chapter assumes only what the chapters before it have already explained. If you start at Chapter 1 and work forward, you will never hit a term that has not been introduced, and you will never need outside reading to follow the argument.

That said, different readers have different goals, and the book accommodates them.

If you are a manager who needs a fast orientation, you can read Chapters 1, 2, 4, 5, 10, and 11 in that order and get a solid working understanding of the technology and its real-world context. You can fill in the middle chapters later if a specific question sends you back to them.

If you are a student building toward more technical study, follow every chapter in order, then work through Appendix B, and finish with Appendix D, which will point you toward the next books and courses that make sense after this one.

If you are an enthusiast who has already read about quantum computing elsewhere, you may find that Chapters 1 and 2 move quickly for you. Feel free to skim them and use the Key Concepts Checklist at the end of each chapter as a self-check before moving on.

Every chapter opens with a two-paragraph Overview that tells you what you are about to read and why it matters. Every chapter closes with a Key Concepts Checklist (a short self-check), four open-ended Questions to Explore, a Try This or Reflect activity, a Glossary Terms section with six definitions, and a five-bullet Key Takeaways list. These closing sections are useful whether you are reading solo, leading a study group, or teaching a class.

The diagrams throughout the book are central, not decorative. They carry ideas that are genuinely easier to show than to say. Each one is described in its caption, but you will get more from pausing and looking than from skipping ahead. Resist the urge to rush past them.

Appendix B sits at the back for readers who want to trace the logic of BB84 with actual numbers. It is clearly marked as optional and written to be self-contained, so you can drop in whenever curiosity strikes without needing to have read any specific prior chapter.

Appendix C lists free resources: simulators you can run in a browser, short videos that animate the protocol, and a few beginner-friendly courses. They are all free, accessible without specialist software, and curated for beginners rather than researchers.

Appendix D gives a reading path for what comes after this book. It is organized by level, annotated with what each resource teaches, and designed so you can take the next step without feeling lost.

One note on vocabulary: this book uses acronyms freely, with their expansions provided on first use. QKD appears hundreds of times after Chapter 1 defines it. The Glossary at the back lists every defined term alphabetically with its plain-language definition, so if you ever land on a term that feels unfamiliar, a quick flip to the back will resolve it.

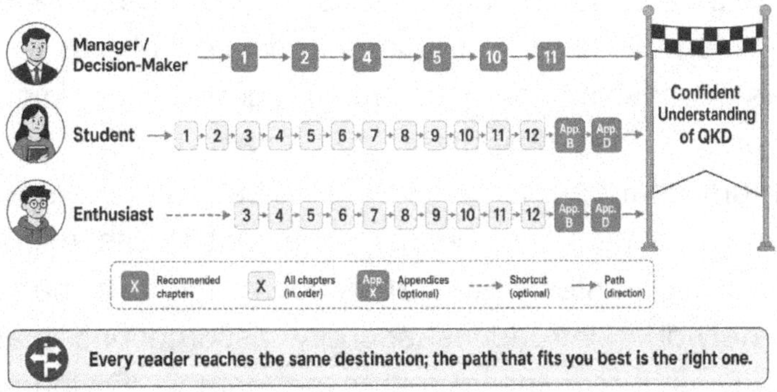

Diagram I.3: Reading paths for different readers

A Promise of Honesty

There is a particular kind of technology book that reads like a press release. It describes a new development as revolutionary, inevitable, and already transforming everything. It focuses only on the best results from the best labs under the best conditions. It mentions competitors only to dismiss them. By the final page, you feel excited but also slightly suspicious, because nothing in your experience of technology has ever been that clean.

This book is not that book.

Quantum key distribution is genuinely remarkable. The physics behind it is beautiful, and the detection guarantee it offers is unlike anything in classical cryptography. Real systems are operating in real deployments right now. The field is advancing quickly. There is every reason to take it seriously.

And yet. QKD has real limitations, and you deserve to know them. It currently requires specialized hardware that is expensive and not easy to integrate into existing networks. The detection guarantee covers the quantum channel but not the classical computers at each end, which can be compromised in the usual ways. The "trusted node" architecture used in many real networks introduces security assumptions that partially undercut the quantum advantage. Distance remains a hard problem: quantum signals in fiber degrade over roughly one hundred kilometers without a repeater, and quantum repeaters that can extend that range without introducing new trust assumptions are still in the research phase.

Post-quantum cryptography, the software-based alternative, has its own profile of strengths and weaknesses. It is much cheaper to deploy and can run on today's hardware, but it remains, at its core, dependent on assumptions about mathematical hardness, just as today's encryption is. The two approaches are not rivals; they are complements, and the most serious security architects treat them as such.

The book takes the position that a reader who understands the limitations of a technology is a better user of that technology than one who has only been shown its highlights. Every chapter that explains what QKD can do is paired, eventually, with an honest account of what it cannot. Chapter 10 is devoted entirely to limitations, and it pulls no punches. This is not pessimism. It is the respect you deserve as a reader.

When the book describes something as proven, it means physics has demonstrated it under repeated, independent experimental conditions. When the book describes something as promising, it means the research is encouraging, but the engineering is not yet mature. When the book describes something as speculative, it means serious scientists are working on it, but no one can promise a timeline. These distinctions matter, and the book maintains them throughout.

Science is a process, not a destination. The quantum internet described in Chapter 12 is a genuine scientific goal, pursued by real researchers at real institutions. It is also not here yet, and it may be a decade or more away in any practical form. Treating that honestly is not a criticism of the researchers. It is fidelity to how science actually works.

The author of this book has a PhD in a relevant technical field and has spent years making difficult ideas accessible to non-specialist audiences. That combination creates one particular obligation: never let the desire to excite the reader outpace the commitment to telling the truth. If something in this book turns out to be wrong, it will be corrected in future editions. If you spot an error, you are encouraged to write in. Good books get better with honest readers.

The promise, then, is this: every claim in this book is as accurate as careful research can make it. Every limitation is as visible as every strength. Every analogy is tested for accuracy before it is trusted to carry an idea. And when the story of quantum security is still unfolding, as it is in several places right now, the book says so rather than pretending the future is already known.

That is the only kind of honesty that earns trust. And trust is what this whole subject is ultimately about.

1 Secrets, Keys, and Why They Matter

Overview

Almost everything private in modern life is protected by encryption, and encryption depends on a secret key. A key is simply a string of numbers that both sides of a conversation share and that no one else knows. With the key, scrambled data becomes readable. Without it, the same data looks like meaningless noise. This chapter builds that mental model carefully, because every later chapter in this book is ultimately about one question: how do two people who have never met agree on a secret key that nobody else can learn?

The chapter introduces the oldest and hardest problem in all of cryptography, which is the key distribution problem, and shows why it has never been fully solved by clever thinking alone. It explains, in everyday terms, the elegant trick that today's internet uses to get around it. It also plants the seed of the book's central tension: that trick relies on certain math problems being hard to solve, and "hard math" is an assumption rather than a law of nature. Assumptions can break. By the end of this chapter, you will understand what a key is, why sharing one securely is so surprisingly difficult, and why a method rooted in physics rather than mathematics would represent a profound change in how the world keeps its secrets.

Opening Scenario

Imagine two friends, Priya and Marcus, separated by an ocean, who want to exchange a private letter. They agree in advance to use a padlock system. Priya mails Marcus a padlock that she has locked shut, keeping the only key. Marcus puts his letter inside a small box, snaps Priya's padlock on it, and mails the box back. Priya opens it with her key. Nobody who handled the box in transit ever had the key so that nobody could read the letter. Clever. But there is a catch that almost everyone misses: before any of this could work, Priya had to get that padlock to Marcus in the first place. She mailed it unlocked through the same postal system she does not fully trust. Anyone along the way could have swapped it for a fake padlock they controlled. Then they could intercept the box, open it with their own key, read the letter, re-lock it with Priya's real padlock, and forward it. Neither friend would ever suspect a thing.

That moment, the moment when the unlocked padlock travels unsupervised through the world, is the key distribution problem. It is not a puzzle with an easy answer. It is the reason mathematicians, engineers, and physicists have spent decades building ever more elaborate systems. The padlock in the story is a metaphor for the digital equivalent of a shared secret. Getting that secret from one place to another without anyone copying it is the central challenge of all secure communication.

This book is the story of how physics offers an answer that math alone cannot guarantee. But to appreciate the answer, you first need to understand the problem clearly, and that is exactly what this chapter does. We will build the picture from the ground up, starting with what encryption actually does and ending with a clear view of why the world is searching for something better.

What This Chapter Will Cover

• What encryption really does, using the everyday image of a locked box and a reversible scramble.

• What a "key" actually is, demystifying the word and separating it from passwords you type.

• The key distribution problem, the central ancient difficulty that every secure channel must face.

• How today's internet solves the problem, using the clever "padlock anyone can snap shut, but only I can open" idea.

• The hidden assumption behind that solution, and why assumptions are not guarantees.

• Symmetric versus public-key encryption in plain terms, and how the two kinds team up every day.

• Why all of this sets up the rest of the book, and where quantum key distribution (QKD) fits in.

What Encryption Really Does

Think of a message as a piece of paper with words on it. Encryption is like feeding that paper through a machine that rearranges every letter according to a secret rule. The output looks like gibberish. Someone who does not know the rule cannot recover the original words, no matter how long they stare at it. Someone who knows the rule runs the machine in reverse and gets the original paper back perfectly. That reversible scrambling is all encryption is. There is no magic, no impenetrable vault: just a rule and a secret.

The "secret rule" is what engineers call the key. The key is a number, usually very long, that controls exactly how the scrambling occurs. Change one digit in the key, and the output scrambles differently. Use the wrong key to unscramble, and you get nonsense. Use the right key, and the message snaps back into readable form instantly. The message does not know or care who it is addressed to. What matters is whether you hold the correct key.

Here is the part that surprises many people: the scrambling algorithm itself, the mechanism of encryption, is usually public knowledge. Researchers publish it. Governments study it. Anyone can read the specification. The security does not come from hiding the method. It comes entirely from hiding the key. This is one of the foundational principles of modern cryptography, and it has a name: Kerckhoffs's principle. The idea is that a system should still be secure even if everything about it is known except the key. Think of a combination lock. Everyone knows how combination locks work. The only secret is your specific combination.

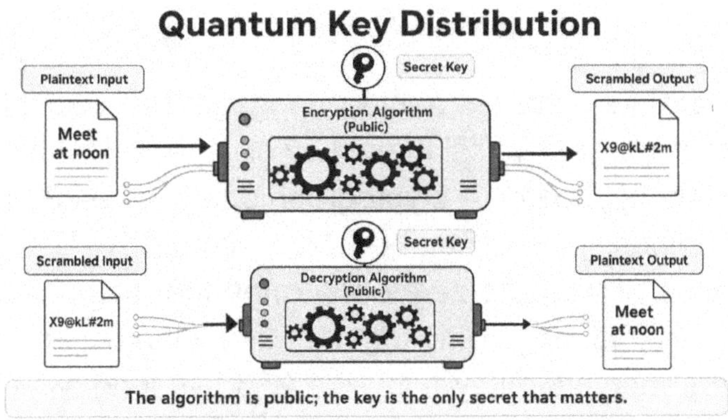

Diagram 1.1: Encryption as a reversible scrambling machine

This means that if you wanted to break a well-designed encrypted message, your best option is to figure out the key. The algorithm alone tells you nothing useful. This is liberating for a strange reason: it means the entire problem of secure communication reduces to one question. Can you get the key to the right person without anyone else learning it? Everything else- the algorithms, the protocols, the padlocks, the math- is scaffolding around that single question.

It also means that a weak key, or a carelessly shared key, can undo the strongest encryption in the world. A hundred-page novel locked in a steel safe is worthless if you tape the key to the outside. Encryption is only as strong as your key management. And key management begins with key distribution: how does the right person get the right key in the first place?

What a "Key" Actually Is

The word "key" carries a lot of baggage. Most people picture either a physical metal key or a password they type on a keyboard. A cryptographic key is neither of those things. It is a number. More precisely, it is a sequence of bits, the zeros and ones that computers use for everything, strung together into a long chain. A modern encryption key might be 256 bits long, which means it is a number with 256 binary digits. Written as a decimal, that number has about 77 digits. Written as random-looking characters on a screen, it is something you would never memorize and would never try to.

You rarely see your own encryption keys. When you visit a website with a padlock icon in the browser's address bar, your computer and that website's server negotiate and automatically create a fresh key behind the scenes in a fraction of a second. The key exists for only that session and then disappears. You did not choose it, you cannot read it, and you do not need to think about it. The software handles it invisibly. That invisibility is a feature, not a bug, because human-chosen keys tend to be terrible. People gravitate toward short, memorable sequences, and those are the first ones an attacker tries.

A password is a different concept. A password is something you type to prove your identity to a system. A key is something two parties share to scramble and unscramble data between them. The two ideas sometimes overlap in practice. For example, a password manager might derive an encryption key from the password you type. But conceptually they are distinct. A password unlocks access. A key transforms data.

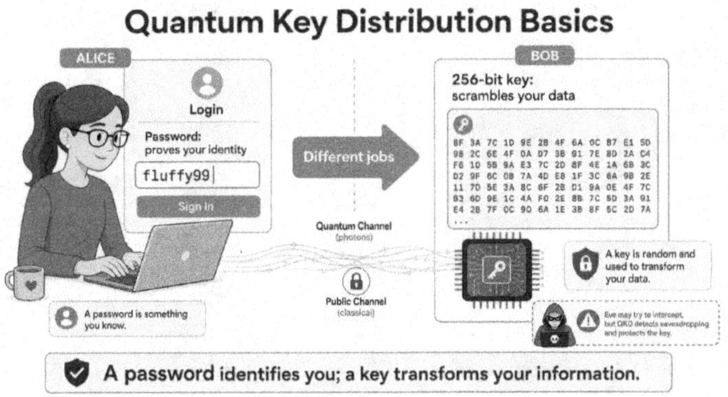

Diagram 1.2: A key as a long number, not a password

The length of the key matters enormously. If a key is only 8 bits long, there are exactly 256 possible keys. An attacker could try all of them in a blink. If the key is 256 bits long, there are more possible keys than there are atoms in the observable universe. Even a computer that tries a trillion keys per second would need far longer than the age of the universe to exhaust all the possibilities. This is why key length is one of the first things security engineers check. Short keys are breakable. Long keys, chosen randomly, are not.

The word "randomly" matters too. A key must be truly random, not just hard to guess. Predictable patterns weaken keys even when they are long. A 256-bit key that follows some hidden pattern might only require testing a few million possibilities instead of the full astronomical number. This is why cryptographic systems use carefully designed random-number generators, sometimes drawing on physical processes such as electronic noise, to produce keys with no exploitable structure.

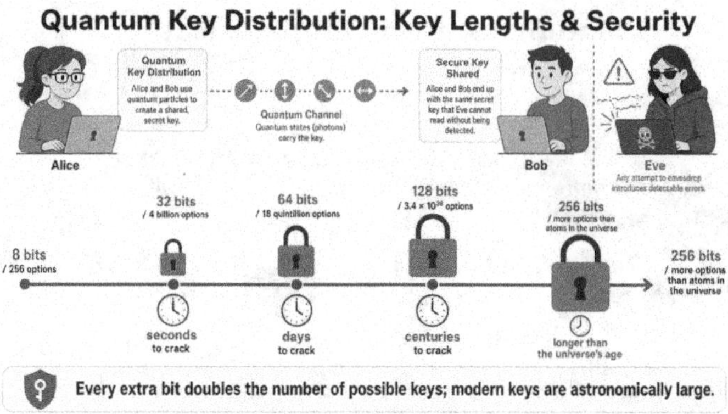

Diagram 1.3: Why key length creates astronomical protection

The Key Distribution Problem

So far, the picture is clean: encryption uses a key; the key must be long and random; and the whole game is about protecting the key. Now comes the uncomfortable part. Before Alice and Bob, the classic characters of cryptography, can exchange a single encrypted message, they must already share the same key. Encryption requires that both sides use the same rule. If Alice scrambles with key number 7 and Bob tries to unscramble with key number 12, he gets noise. They must agree on the key before communication begins.

How do they agree? They could meet in person. Two heads of state could send couriers carrying identical key material in locked briefcases. Spy agencies did exactly this for decades: physically delivering key tapes to embassies, submarines, and field offices around the world. It worked, but it was slow, expensive, and vulnerable to the human beings doing the carrying. A courier can be bribed, copied, photographed, or observed.

For two people who have never met and never will, physical courier delivery is not an option. Imagine a customer in Seoul who wants to log in to a bank server in Toronto securely. They cannot exchange keys in person. They must use the same internet connection they are trying to make secure. But sending the key over that connection seems circular. If the connection is not yet secure, anyone listening can copy the key as it travels, and then use that copy to decrypt every message that follows.

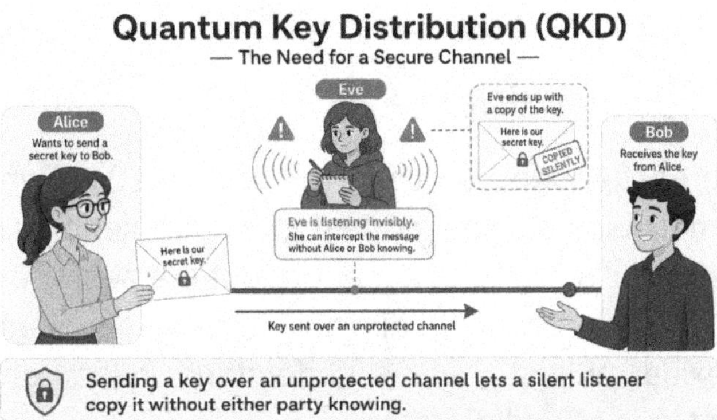

Diagram 1.4: The key distribution problem in action

This is the key distribution problem, and it is ancient. Julius Caesar wrestled with a version of it. World War II was shaped by it. The Allies' ability to break German military codes depended not just on mathematical insight but on captured key material, physical documents that revealed the daily settings of encryption machines. The Germans changed their keys regularly precisely because they understood that captured key material was catastrophic.

In the digital age, the problem multiplied. A single online store might serve millions of customers per day, each one needing a fresh, unique shared key established in real time, without any prior contact. Physical couriers were laughably impractical. Something fundamentally new was needed.

For centuries, mathematicians assumed the problem was unsolvable: if someone could eavesdrop on every channel you used, you could not share a secret with a stranger. That assumption turned out to be wrong, but overturning it required one of the cleverest ideas in the history of mathematics.

How Today's Internet Solves It

The breakthrough arrived in the 1970s, when mathematicians discovered that it is possible to share a secret over a public channel without the secret ever crossing that channel directly. The idea seems paradoxical at first. Let us understand it through a different kind of padlock story.

Imagine that instead of sharing a key with someone, you mail them an open, unlocked padlock. You keep the key to that padlock. Anyone in the world can pick up the padlock and snap it shut around something. But only you can open it, because only you have the key. If your friend in Tokyo wants to send you a private message, she puts the message in a box, snaps your padlock shut, and mails it. Even if the box is intercepted, nobody can open it without your key, which never left your possession. You did not need to share anything secret to receive a private message. You only needed to share an open padlock, something that anyone can have and that carries no secret information itself.

Diagram 1.5: The open-padlock analogy for public-key cryptography

In mathematics, the open padlock represents the public key, and the key that opens it represents the private key. Every participant generates a pair: one key that is publicly published (the public key) and one key that is kept absolutely secret (the private key). Anyone can encrypt a message using your public key. Only you can decrypt it using your private key. The two keys are mathematically linked, but knowing the public key does not tell you the private key.

This system is called public-key cryptography, also known as asymmetric encryption because the two sides use different keys. It solves the key distribution problem beautifully: you can broadcast your public key to the entire planet without risk, because possessing it only allows others to send you private messages. It prevents them from reading your messages or impersonating you.

The browser padlock you see when visiting a secure website reflects this process at work. Your browser and the website's server run through an automated negotiation called a handshake. During that handshake, public-key cryptography is used to agree on a temporary shared secret securely. From that shared secret, both sides derive an identical encryption key that exists only for that session. Nobody who watched all the network traffic could reconstruct that key from what they saw, because the clever mathematics of the handshake prevents it.

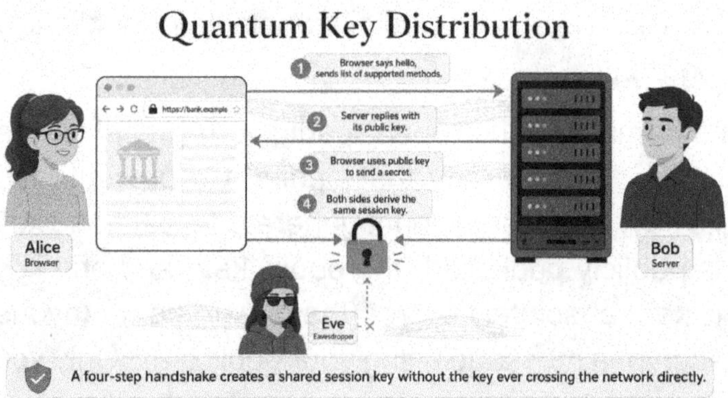

Diagram 1.6: The browser handshake and the padlock icon

The result is remarkable. Two computers on opposite sides of the world that have never communicated before can establish a private channel in milliseconds. No couriers needed. No prior meeting. No shared secret that had to be delivered in advance. The system works because of the one-way nature of the mathematical relationship between the public and private keys: easy to go one way, ferociously hard to reverse.

The Hidden Assumption

Here is where the story takes a turn that most people never hear. Everything in the previous section works, and it works reliably in practice today. But it rests on a specific assumption: that certain mathematical problems are too hard to solve quickly.

The most famous example involves large numbers and their prime factors. Take two large prime numbers, multiply them together, and you get a very large product. This is easy. The multiplication takes a fraction of a second. But run the process in reverse, start with the large product and try to figure out which two primes were multiplied, and the task becomes fiendishly difficult. With numbers large enough, even the fastest computers today would take millions of years to find the answer by brute force.

Public-key cryptography is built on that difficulty. The private key is hidden inside a mathematical relationship whose reverse direction is computationally prohibitive. As long as nobody finds a fast way to solve these "hard" problems, the system is secure.

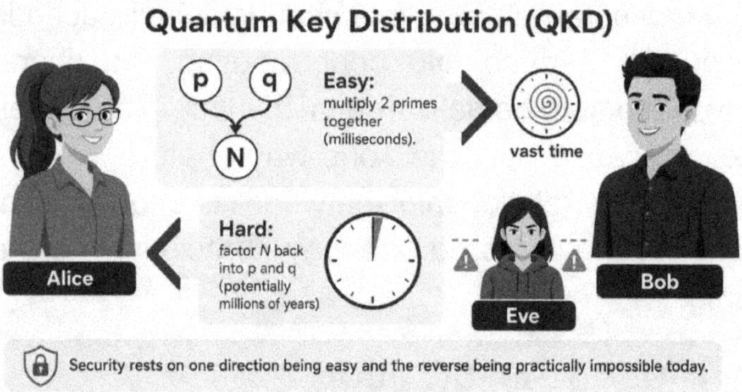

Diagram 1.7: The easy direction versus the hard direction

The word "today" carries all the weight in that last sentence. The assumption is not that factoring large numbers is mathematically impossible. It is that no known algorithm can do it fast enough on current hardware. That is an empirical statement about the state of knowledge at a particular moment in history. It is not a theorem. It has never been proven that a fast factoring algorithm cannot exist. It is simply that nobody has found one yet.

This distinction matters enormously. The entire security of internet banking, private messaging, government communications, and medical records depends on a gap in human mathematical knowledge persisting indefinitely. If tomorrow morning a mathematician published a fast factoring algorithm, or if a new type of computing machine made previously hard problems easy, the foundation of internet security would shift overnight.

That new type of computing machine exists in early, limited form today. It is the quantum computer. Quantum computers use principles of physics that are fundamentally different from the principles governing ordinary computers, and for certain classes of problems, they are dramatically faster. One of those problem classes includes exactly the kind of mathematical puzzles that public-key cryptography relies on. This is why the world is paying close attention to quantum computing, and this is why this book exists.

Diagram 1.8: An assumption is not a law of nature

It is worth pausing here to be precise about the risk. Quantum computers powerful enough to break today's public-key cryptography do not exist yet. Current machines are small and error-prone. The timeline for when (or whether) large-scale quantum computers will arrive is genuinely uncertain among experts. But the preparation has already started, because encrypting data that must stay secret for decades means you have to worry about future code-breaking machines today. Data encrypted now can be harvested and stored by adversaries, who will wait until a powerful quantum computer exists to decrypt it. This is sometimes called the harvest-now, decrypt-later threat, and it is one of the main reasons governments and standards bodies are urgently working on alternatives.

Symmetric vs. Public Keys, in Plain Terms

You have now met two different types of encryption, and keeping them straight will help you follow every chapter that follows. The first type is called symmetric encryption. In symmetric encryption, both sender and receiver use the same key. Think of a combination safe that two business partners share: both know the combination, both can open the safe, and neither has a special advantage over the other. Symmetric encryption is fast and efficient. Modern computers can encrypt vast amounts of data symmetrically in a blink. The challenge, as you have seen, is the distribution problem: how do both parties come to hold the same key without a third party copying it?

The second type is public-key cryptography, which uses two mathematically related but distinct keys: a public key and a private key. This solves the distribution problem beautifully because the public key carries no secret. Anyone can have it. But public-key systems are computationally heavier than symmetric systems. They are slower and less suited to encrypting large volumes of data.

Diagram 1.9: Symmetric and asymmetric encryption side by side

In practice, the internet uses both types together in a layered approach that plays to each one's strengths. Public-key cryptography handles the introduction: two parties who have never met use their public and private keys to negotiate a fresh symmetric key. Once that symmetric key is shared, both sides switch over to symmetric encryption for the actual data. The bulk of your data- your messages, your video, your files- travels symmetrically. The asymmetric layer is only used for the handshake at the beginning.

Think of it as two stages. Stage one: the asymmetric system acts like an armored courier service, safely delivering a sealed envelope containing a fresh key to both parties. Stage two: both parties open the envelopes, pull out the identical symmetric key, and use it for all future communication. The courier stage is expensive and slow but necessary. The symmetric stage is fast and efficient. Together, they cover each other's weaknesses.

This is what happens every time you visit a secure website. The padlock icon means the handshake was completed successfully. Your browser and the server agreed on a symmetric session key using asymmetric cryptography, and now everything between you and that server is encrypted symmetrically. The entire process took milliseconds and required no human involvement. It is one of the quiet engineering marvels of modern life.

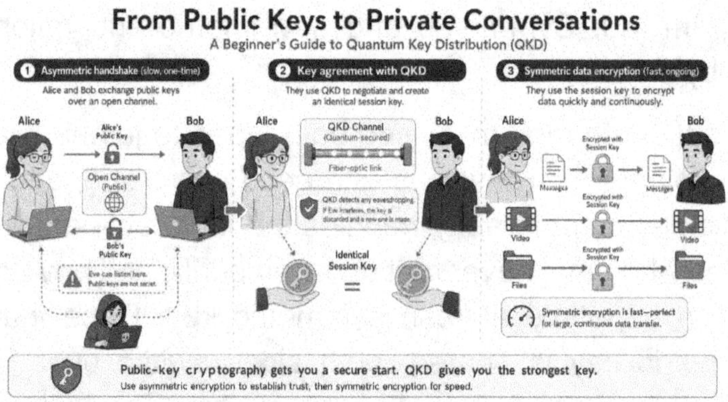

Diagram 1.10: How symmetric and asymmetric encryption work as a team

Understanding this two-stage architecture matters because it shows you where the vulnerability lies. The asymmetric handshake is the exposed moment. If the mathematics underlying the asymmetric system were broken, an attacker who recorded the handshake traffic could reconstruct the symmetric session key and decrypt everything. The whole session would unravel from its first moment. This is not a theoretical concern invented by researchers; it is the specific scenario that quantum computers could eventually make real.

Why This Sets Up Everything That Follows

Now you can see the full picture, and the tension within it. Modern encryption is brilliantly clever. It protects billions of transactions every day. It works beautifully under current conditions. But it rests on a foundation with a known potential weakness: the assumption that certain math problems will remain hard as computing technology evolves.

The question the rest of this book addresses is straightforward: what do you do if that assumption breaks? There are two broad families of answers being developed simultaneously, and understanding both is essential context for a thoughtful person following technology news today.

The first family is called post-quantum cryptography (PQC). The idea is to replace the current math-based public-key systems with different mathematical problems that quantum computers cannot solve efficiently. Researchers have been working on this for years, and standards bodies have recently selected the first batch of post-quantum algorithms for official use. PQC is a software upgrade: you swap out the cryptographic system's mathematical core while keeping everything else the same.

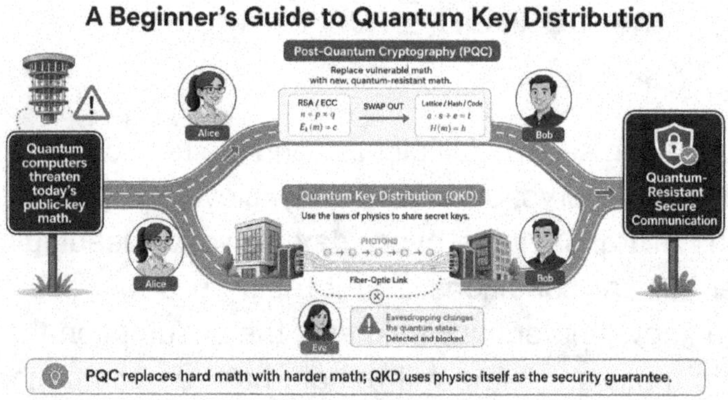

Diagram 1.11: Two responses to the quantum threat

The second family is quantum key distribution (QKD). Instead of finding better math, QKD uses the laws of quantum physics to distribute encryption keys in a way that makes eavesdropping physically detectable. The security guarantee does not come from a mathematical assumption. It comes from properties of the universe that clever algorithms or more powerful computers cannot change. You cannot improve a computer to get around the laws of quantum mechanics, because quantum mechanics is a description of how the universe actually behaves.

These two approaches are not mutually exclusive. Many security architects argue that a truly robust post-quantum system should use both: PQC to handle the mathematical layer and QKD to protect the key distribution process with physical guarantees. The two strategies address overlapping, but slightly different problems, and their combination offers defense-in-depth.

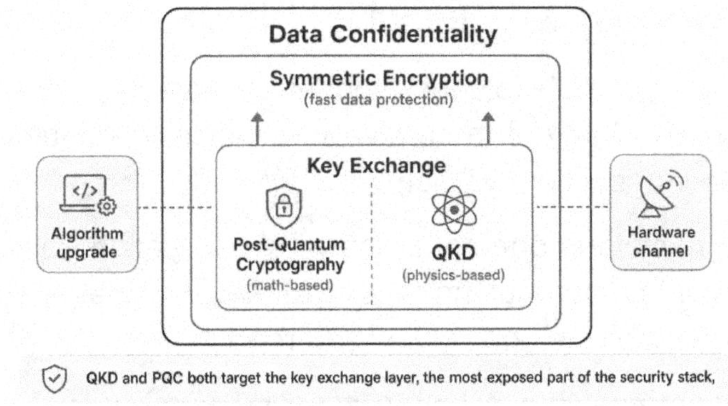

Diagram 1.12: Where QKD fits in the broader security picture

The rest of this book zooms in on QKD: how it works, what makes it physically secure, where it is already deployed, and what challenges remain before it can be used as widely as the internet itself. Along the way, you will learn the quantum physics you need, explained without any equations, through stories and analogies. You will see real networks, real challenges, and a realistic assessment of both the promise and the limits of this technology.

Before you can appreciate QKD's elegance, however, you need to understand the physical world it operates in. That is where the next chapter begins: with light, which is the medium QKD uses to carry its keys. But everything in those later chapters stands on the foundation you built here. Keys are secret numbers. Distribution is the hard part. Today's solution rests on an assumption. Physics may offer something more solid. That progression is the spine of this entire book.

Key Concepts Checklist

Use these prompts to check your understanding before moving on. If any of them give you pause, re-reading the relevant section will help.

• I can explain in one sentence what encryption does: it scrambles data using a secret key so that only someone with the correct key can read it.

• I understand that a key is a secret number, not a password I choose, and that longer, random keys are exponentially harder to crack.

• I can describe the key distribution problem to a friend: how do you get a secret key to someone without anyone else copying it on the way?

• I know why the browser padlock relies on a math problem being hard: public-key cryptography uses one-way mathematical functions whose reversal is computationally prohibitive on current machines.

• I understand that "hard math" is an assumption about computational difficulty, not a proven impossibility, and that new techniques can invalidate assumptions.

• I can tell the difference between symmetric encryption (same key on both sides, fast) and asymmetric or public-key encryption (different keys, solves distribution, slower).

• I understand how the two types work together: asymmetric for the handshake, symmetric for the bulk data.

Questions to Explore

These questions have no single right answer. They are designed to deepen your thinking, spark good conversations, and help you connect the ideas here to your own experience.

• If you had to send a house key to a friend across the country safely, without meeting in person, what methods could you think of? What makes each method vulnerable?

• Why might it be dangerous for the entire internet to depend on a single assumption about how hard a math problem is? What does that dependence mean for long-term data security?

• Can you think of a secret you share with someone that no third party knows? How did you both come to know it, and how confident are you that nobody else learned it along the way?

• What would change about daily life if encryption suddenly stopped working? Which parts of your day would be most immediately affected?

Try This / Reflect

Pick up a pen and try the classic three-padlock thought experiment with a friend. Each of you imagines owning a box you can lock, but only you can unlock it. The challenge: send a private note using only boxes you each control, without ever sharing keys. You are not allowed to meet in person, and any box may be copied or inspected in transit. Think through the steps before checking the answer. When you are ready, search online for a short animated video titled "public key cryptography explained" and watch how close your solution was to the real method. Pay attention to the moment when the video resolves the apparent paradox. That moment of resolution is the insight that changed the world in the 1970s, and it will feel satisfying once you see it.

If you prefer a quieter reflection, think about the last time you logged into a website on your phone. In the second or two it took for the page to load, your device and that server negotiated a fresh encryption key using asymmetric cryptography, then switched to symmetric encryption for everything that followed. You did not lift a finger. How does it feel to know that a decades-old mathematical insight was quietly protecting you in that moment?

Glossary Terms

• **Encryption** - The process of scrambling information using a key so that only someone possessing the correct key can restore it to readable form.

• **Key** - A secret number, typically a long random sequence of bits, shared between a sender and receiver to control how data is scrambled and unscrambled.

• **Key distribution problem** - The fundamental challenge of sharing a secret key over a channel that may be monitored, without allowing any observer to copy the key.

• **Public-key cryptography** - An encryption method that uses a mathematically linked pair of keys (one public, one private) so that two parties can establish a shared secret without meeting in advance.

• **Symmetric encryption** - An encryption approach in which both parties use the same key to scramble and unscramble data; fast and efficient but dependent on solving the key distribution problem first.

• **Post-quantum cryptography** - A family of mathematical encryption methods designed to remain secure even against attacks from quantum computers, by using problems that quantum algorithms cannot solve efficiently.

Key Takeaways

• All digital privacy ultimately depends on a secret key, and every security system must solve the problem of getting that key to the right person without anyone else copying it.

• The key distribution problem is ancient and was long considered fundamentally unsolvable for strangers communicating over an untrusted channel.

• Today's internet solves it with public-key cryptography, an elegant mathematical trick that lets two parties establish a shared secret without the secret ever crossing the open channel directly.

• That solution rests on the assumption that certain math problems are too hard for current computers to solve quickly. Assumptions can be overturned, and quantum computers specifically threaten this one.

• Post-quantum cryptography and quantum key distribution are the two leading responses to that threat: one replaces the math with harder math, the other replaces math-based security with security grounded in the laws of physics.

2 The Quantum Threat and Harvest Now, Decrypt Later

Overview

For most of the digital age, the locks protecting your data have relied on a simple idea: certain math problems are so hard that no computer could crack them in any reasonable amount of time. A problem that would take a billion years to solve by brute force is, for all practical purposes, unsolvable. That hardness is the foundation beneath your bank's website, your medical portal, your private messages, and virtually every encrypted channel on the internet. This chapter explains, in plain language and without a single equation, why a new kind of machine threatens to shatter that foundation. You do not need to understand how a quantum computer works at the circuit level. You need only to understand what it would mean if one arrived.

The second half of this chapter deals with something that surprises almost everyone: the threat does not wait for quantum computers to exist. Adversaries with enough patience and storage can record your encrypted data today and wait. When a powerful enough quantum computer eventually appears, they decrypt the old traffic in bulk. Every secret inside that traffic, no matter how old, becomes visible. This strategy has a name, and by the end of this chapter you will understand it well enough to explain it to someone else. You will also have a practical tool for judging how urgently any given secret needs protection. That urgency is what drives every government standard, every industry migration, and every technology described in the chapters ahead.

Opening Scenario

Picture a server room somewhere in a large city. Racks of blinking machines hum as they route data for two banks that move money between each other thousands of times a day. Every packet of that data is wrapped in encryption so strong that intercepting it and reading it would take, by current estimates, far longer than the age of the universe. A security officer at one of the banks checks the system logs and sees everything is normal. She goes home satisfied.

Somewhere else, on hardware that costs very little, a copy of that same traffic is being saved. Not decoded, not analyzed, just saved. An adversary knows they cannot read it today. They are not even trying. They are simply building a library, disk by disk, filing cabinet by filing cabinet, waiting for a tool that does not yet exist in a form powerful enough to be useful. They do not need to understand every transaction inside the encrypted stream. They only need to know that one day, when the right machine exists, the entire archive will open.

The security officer believes the bank's communications are safe. In a narrow sense, she is right. But the copies sitting in that quiet archive were made years ago. The moment a sufficiently powerful quantum computer comes online, those copies become readable. The secrets inside them- contracts, account balances, regulatory filings- will become exposed at a moment the bank cannot predict and against a threat it may never have cataloged. The lock on today's data is strong. The problem is that someone is already holding the sealed box, waiting for the key.

What This Chapter Will Cover

• What makes a quantum computer fundamentally different from an ordinary computer, explained with a maze analogy and no math.

• The specific algorithm that could crack the public-key encryption protecting most of the internet, and why you only need to grasp its consequence, not its inner workings.

• Which kinds of encryption are seriously threatened and which remain largely safe so that you can avoid both under-reaction and over-panic.

• An honest survey of where quantum computers stand today and why uncertainty about timing is itself a reason to act rather than a reason to wait.

• The "harvest now, decrypt later" strategy in full detail, including who is most at risk and why.

• The shelf life of a secret, a simple concept that lets you estimate how urgently any particular piece of information needs protection.

• Why this threat creates real demand for quantum key distribution (QKD), and how that sets up everything in the chapters ahead.

What Makes a Quantum Computer Different

To understand why a quantum computer threatens today's encryption, you first need a clear picture of what makes it different from the laptop or phone you use every day. The difference is not merely speed. It is a fundamentally different way of approaching certain problems.

An ordinary computer, whether it is a phone, a desktop, or a supercomputer filling an entire building, works by processing information as bits. Each bit is either a zero or a one, off or on, like a light switch. Every calculation the machine performs moves through these zeros and ones one step at a time. If you ask it to find the way through a maze, it picks one path, follows it until it hits a wall, backs up, tries another path, and repeats. A faster computer does this cycling more quickly. The approach is the same.

A quantum computer uses a different kind of unit called a quantum bit, or qubit. Thanks to the rules of quantum physics, a qubit can exist in a condition called superposition before it is measured. Think of it not as a coin lying flat showing heads or tails, but as a coin spinning in the air, holding both possibilities at once until it lands. While the qubit is in superposition, a quantum computer can perform operations that take both possibilities into account simultaneously. When you link many qubits together, the number of simultaneous possibilities grows exponentially. Ten qubits can represent over a thousand states at once. Three hundred qubits can represent more states simultaneously than there are atoms in the observable universe.

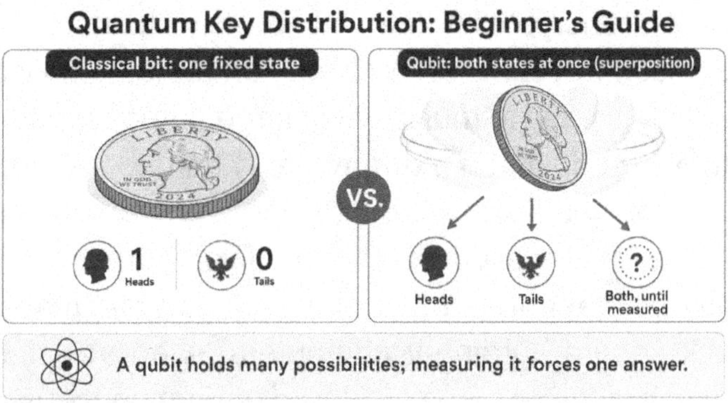

Diagram 2.1: Classical bits versus qubits

Now, superposition alone does not make quantum computers magic. The real power comes from another quantum property called interference. A quantum computer can be designed so that wrong paths cancel each other out (like two waves canceling in water) and correct paths reinforce each other. The result is that the right answer rises to the surface while the wrong ones fade away. For most everyday computing tasks, this provides no benefit. For a narrow class of mathematical problems, however, it provides an advantage so enormous it breaks the problem open entirely.

The most famous of those problems is the one behind public-key encryption. To understand the threat, you need to see why that particular problem is the one that quantum mechanics happens to solve so well.

The Algorithm That Breaks Today's Locks

Public-key encryption, the technology behind the padlock icon in your browser, rests on a mathematical trapdoor. It is easy to multiply two large prime numbers to obtain a product. It is extraordinarily hard to take that product and work backward to find the two original primes, a task called factoring. The difference in difficulty between those two directions is what makes the lock secure. Ordinary computers, even very fast ones, cannot factor a number large enough to be used in modern encryption within any practical amount of time.

In 1994, a mathematician named Peter Shor discovered a procedure for quantum computers that makes factoring dramatically faster. You do not need to know how Shor's algorithm works at a technical level. What matters is the consequence: when run on a sufficiently large and reliable quantum computer, it can factor the numbers underpinning today's public-key encryption in hours or days rather than billions of years. The lock is the same. The key-making process, however, becomes trivial.

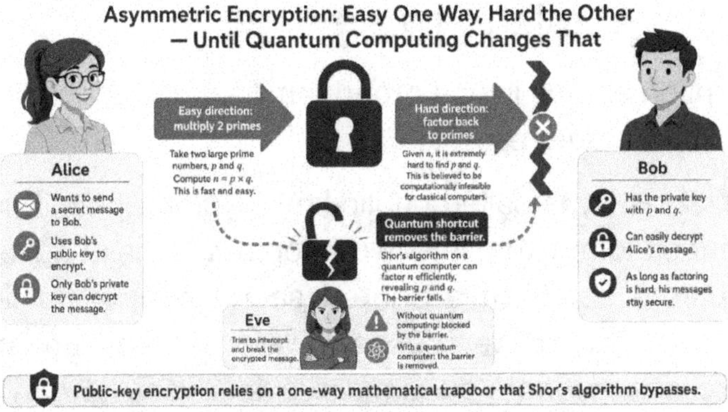

Diagram 2.2: The factoring trapdoor

Shor's algorithm does not attack every kind of encryption equally. It specifically targets the family of problems that public-key systems are built on, primarily factoring large integers and a related problem called the discrete logarithm. These are the mathematical foundations of RSA encryption (the basis of most HTTPS connections), elliptic-curve cryptography (used in modern mobile and financial systems), and the Diffie-Hellman key exchange (used to establish shared secrets across untrusted networks). If a large quantum computer existed today, all three of those technologies would be effectively broken.

It is worth pausing here because the phrase "Shor's algorithm breaks encryption" sometimes causes people to imagine that all digital security collapses overnight. That is not accurate. The algorithm is powerful but narrow. To understand what is and is not threatened, we need to look at how encryption actually divides into two quite different families.

What Is and Isn't Threatened

Encryption comes in two broad families, and they face very different futures.

The first family is called public-key cryptography. This is the technology that lets two people who have never met before establish a shared secret over a public network. Your browser uses it when it sets up an HTTPS connection. Your email provider uses it when it exchanges session keys. This family relies entirely on mathematical hardness, and Shor's algorithm attacks exactly that hardness. Public-key cryptography is genuinely threatened.

The second family is called symmetric encryption. In symmetric encryption, both parties already share a secret key, which is used to encrypt and decrypt the data. The most widely deployed example is the Advanced Encryption Standard (AES), which protects everything from Wi-Fi traffic to hard drives to classified government files. Symmetric encryption does not rely on factoring or discrete logarithms. Quantum computers can attack it using a different approach (a procedure called Grover's algorithm), but that attack is far weaker. Grover's algorithm roughly halves the effective key length. That means a 256-bit AES key, which would take impossibly long to crack with classical computers, becomes roughly as hard to crack as a 128-bit key against a quantum attacker. The defense is simple: use longer keys. A 256-bit symmetric key, already widely used today, is considered quantum-safe, with no structural changes required.

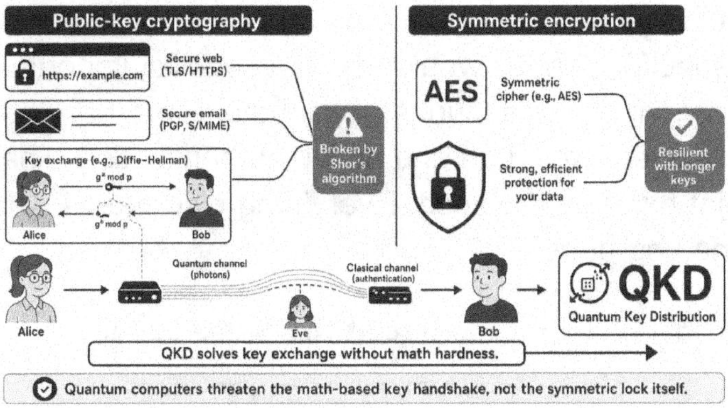

Diagram 2.3: Two families of encryption and their quantum risk

This distinction matters enormously. The popular narrative that "quantum computers will break all encryption" is misleading. What quantum computers threaten is the way we currently share and establish keys, not the symmetric locks that actually scramble the data. Once two parties have a shared symmetric key, the data it protects can be made quantum-safe. The hard problem is getting that key to both parties without exposing it. That is the key distribution problem, and it is exactly the gap that quantum key distribution is designed to fill.

There is a third category worth mentioning: cryptographic hash functions, used for digital signatures and data integrity checks. Grover's algorithm weakens these somewhat as well, but the response is the same as with symmetric encryption: use larger hash outputs. The cryptographic community already offers quantum-resilient hash sizes, and the transition there is manageable.

The picture, then, is not one of total collapse. It is one of a specific, serious weakness in a specific, important component of our security infrastructure, the part that handles key exchange and public authentication. That is what must be replaced. Everything else needs only routine upgrades.

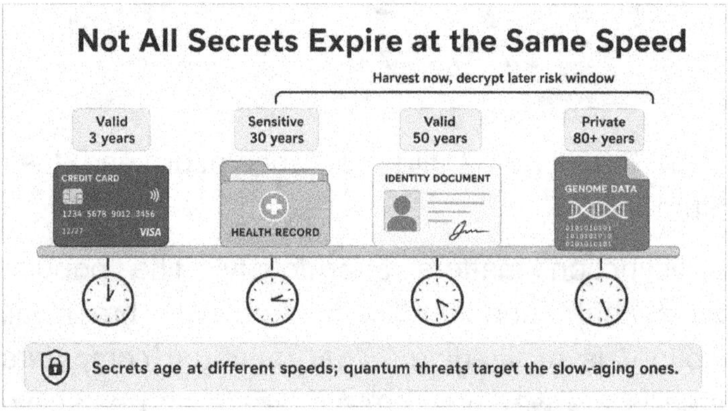

Diagram 2.4: The shelf life of a secret

How Far Away Is the Threat

The honest answer is that nobody knows exactly when a large, reliable quantum computer capable of running Shor's algorithm against real encryption will exist. Estimates from serious researchers range from as little as ten years to as many as thirty or more. Some researchers believe we are closer than official projections suggest; others believe the engineering challenges are so severe that the most threatening machines are still decades away. Anyone who gives you a confident specific date is guessing.

What does exist today is a rapidly growing collection of smaller, noisier machines. These systems, often called NISQ devices (Noisy Intermediate-Scale Quantum computers), can perform certain quantum operations and have already proven useful in research settings. They are not capable of running Shor's algorithm at a scale that would threaten modern encryption. The numbers involved in real-world encryption require millions of stable, error-corrected logical qubits. Current state-of-the-art machines operate with hundreds to a few thousand physical qubits, most of which are too noisy to support the long calculations required for cryptographically relevant attacks.

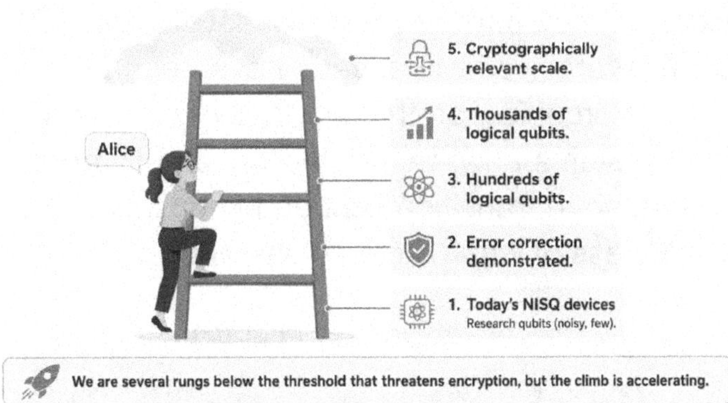

Diagram 2.5: The quantum maturity ladder

The path from today's machines to cryptographically relevant ones requires solving the problem of quantum error correction. Quantum states are fragile. The slightest interaction with the environment- a vibration, a stray electromagnetic field, even the act of measuring- can destroy the superposition a calculation depends on. To run a long algorithm reliably, engineers must use many physical qubits to represent each logical qubit, constantly checking for and correcting errors. The best current estimates suggest that one useful logical qubit may require anywhere from hundreds to thousands of physical qubits for adequate error correction. That gap is closing, but it has not closed yet.

Why does uncertainty matter strategically? Because the cost of acting too late is catastrophic and irreversible, while the cost of acting too early is mostly wasted effort. If an organization waits until a cryptographically relevant quantum computer clearly exists before upgrading its security, the window to protect its legacy data will already have closed. The harvest-now archives will already have been built. By contrast, an organization that starts migrating its most sensitive systems now, before the threat fully arrives, pays a moderate transition cost and keeps its secrets safe regardless of when the machine appears.

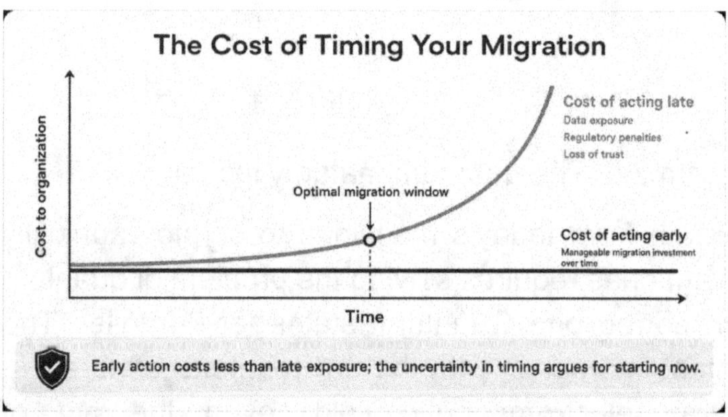

Diagram 2.6: The uncertainty window and decision cost

There is a second, more subtle reason uncertainty is itself a call to action. When you do not know whether a threat arrives in five years or twenty-five, you have to plan for the worst reasonable case, not the average. A five-year horizon means that systems requiring two years to migrate and three years of exposure protection would need to start immediately. Even if the true arrival is fifteen years away, no harm is done by having already finished the migration. The asymmetry between too early and too late strongly favors early action.

Harvest Now, Decrypt Later

The harvest now, decrypt later strategy is the reason this chapter exists. Without it, the quantum threat would be purely a future problem, something to plan for gradually over many years. With it, the threat is already partially present.

Here is how the strategy works. Today, organizations send data over the internet wrapped in public-key encryption. An adversary who intercepts that data cannot read it. The encryption is strong, and the adversary lacks the computational power to crack it. So they do something simpler: they record it. They store the encrypted traffic on cheap hard drives, filling warehouses if necessary, and wait. When a sufficiently powerful quantum computer eventually exists, they load up those old recordings and run decryption in bulk. Every secret that was ever transmitted in that traffic, no matter how old it is, becomes readable at that moment.

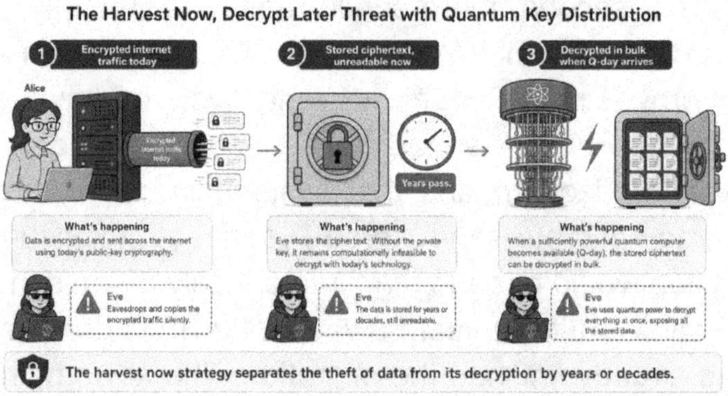

Diagram 2.7: The harvest now, decrypt later pipeline

The strategy works because the encryption protecting most internet traffic today is designed to be strong against today's computers. It was never designed to resist an adversary who could afford to wait and eventually gain access to a fundamentally different kind of machine. The assumption underlying public-key encryption is that factoring large numbers is hard. That assumption is not wrong for today's computers. It is not guaranteed to remain true forever.

Who runs this kind of operation? Nation-states with large intelligence budgets are the most obvious candidates. The cost of storing enormous amounts of data has dropped dramatically over the past two decades. A petabyte of storage, enough to hold years of high-volume network traffic, costs a few tens of thousands of dollars at commodity prices. For a state with strategic interests in the long-term contents of another nation's encrypted communications, this is a trivially affordable investment. Researchers who study intelligence operations have publicly noted that such harvesting operations are almost certainly already underway.

It is worth being precise about what the adversary captures and what they skip. They are not interested in your shopping cart or today's weather query. Those are low-value, short-lived. What makes the strategy dangerous is its value for high-stakes, long-lived information: diplomatic cables, defense contracts, proprietary research, health records, legal proceedings, financial agreements, and intelligence sources. If any of those data categories pass through public-key-encrypted channels today and those channels are recorded, the contents will eventually be exposed.

Diagram 2.8: High-value targets for harvest now archives

The defense against harvest now, decrypt later is not to wait for better computers on your side. The defense is to ensure that what is collected cannot be decrypted even by a quantum computer. That means moving to encryption schemes that are not based on factoring or discrete logarithms, a field called post-quantum cryptography (PQC). It also means that, for the most sensitive communications, a key-sharing method is secure by the laws of physics rather than by mathematical hardness. That second option is what quantum key distribution offers.

The Shelf Life of a Secret

Different secrets have different lifespans, and that difference determines how urgently each one needs quantum-safe protection. This is a simple idea, but it is one of the most practically useful frameworks in this entire book.

Think about the range of information an organization might protect. A password for a one-time promotional discount code might need to stay secret for a few hours. A credit card number needs to stay private for the life of the card, typically three to five years. A health record might contain information the patient would want kept private for decades. A national intelligence source, if identified, could cause harm for the lifetime of the people involved. A genome sequence is essentially private forever because a person's genetic information cannot be changed.

Now layer the harvest-now threat on top of that variation. An adversary who records your traffic today will be able to decrypt it in some number of years. Call that window the quantum arrival uncertainty, which today spans roughly ten to thirty years. If your secret only needs to stay private for two years, the harvest-now threat barely applies to you: by the time a quantum computer exists, your secret will long since have expired or been changed. But if your secret needs to stay private for twenty years, and the quantum arrival window overlaps with that requirement, then the threat is real and present.

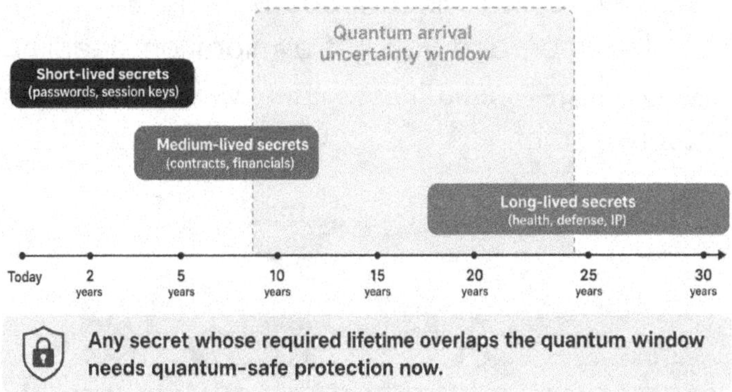

Diagram 2.9: Secret lifetime versus quantum arrival window

The shelf life framework gives you a practical two-part test. First, ask: how long must this information stay secret? Second, ask: could a quantum computer arrive within that window? If the answer to both questions is yes, that information requires quantum-safe protection today, not eventually. This is not a theoretical exercise. Organizations that manage health data, defense information, long-term financial agreements, and intellectual property with multi-decade value need to be applying this test right now.

There is a second component to the shelf life calculation: migration time. Switching an organization's security infrastructure to quantum-safe methods takes time. Inventorying all systems, testing new cryptographic standards, updating software, retraining staff, and completing rollout might take two to five years for a large organization. If the quantum window might open in ten years and migration takes five years, planning needs to start within the next five years to complete before the window opens. Add those two numbers together, and you get a migration deadline. For many large institutions, that deadline is not comfortably distant.

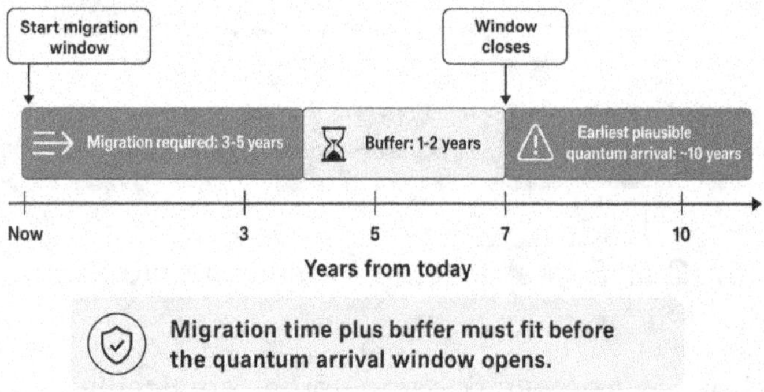

Diagram 2.10: The migration deadline calculation

The shelf life framework also helps communicate urgency without causing panic. You are not telling someone that all their data is in danger. You are asking a precise question about specific data and its specific timeline. That precision tends to produce action rather than paralysis. When a chief information security officer recognizes that the organization's 10-year contracts and research archives require attention. At the same time, its daily transaction logs do not, she can allocate resources appropriately rather than treating the entire problem as either catastrophic or irrelevant.

Why This Creates Demand for QKD

The harvest-now, decrypt-later threat, combined with an honest assessment of where quantum computers are heading, creates a very specific need. There is no need for faster computers, better antivirus software, or tighter firewalls. It is a need for a different way to share cryptographic keys, one that does not depend on any mathematical problem remaining hard forever.

This is where the story of quantum key distribution enters. QKD does not encrypt data with math. It does not rely on the assumption that factoring large numbers is hard. Instead, it uses the laws of quantum physics to distribute a shared secret key between two parties in a way that is, in principle, immune to any computational attack, whether classical or quantum. If someone intercepts the key while it is in transit, the laws of physics make that interception detectable. You cannot silently copy a quantum state without disturbing it. That principle, which you will understand deeply by the end of this book, is what makes QKD structurally immune to harvest-now attacks.

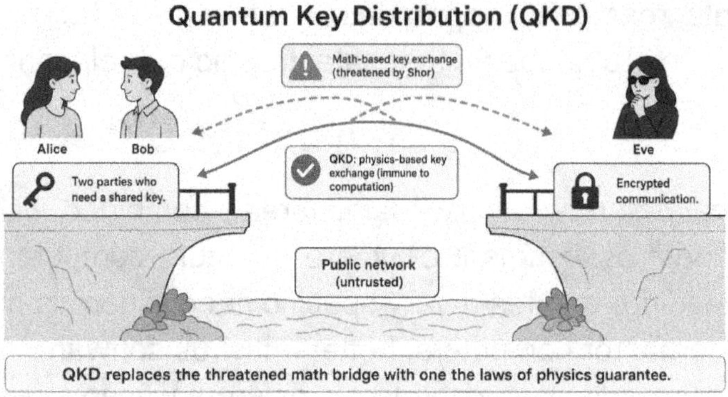

Diagram 2.11: Why QKD fills the gap

Post-quantum cryptography, the other major response to the quantum threat, takes a different approach. Instead of replacing the physics of key exchange, it replaces the math behind it. Researchers have developed new mathematical problems that are believed to be hard even for quantum computers, and standards bodies have been developing and vetting these alternatives for years. Post-quantum cryptography (PQC) is a critically important field and a practical solution for most of the internet's needs. The choice between PQC and QKD is not a competition with a single winner. They are complementary tools, and the chapter on hybrid security later in the book will address when each one fits.

For now, what matters is understanding why QKD exists and why its moment has arrived. Decades ago, researchers developed QKD as a theoretical curiosity, a beautiful demonstration of what quantum physics could do for security. It was genuinely impressive science, but it addressed a problem that most security engineers did not feel was urgent. The math was holding up fine. Nobody was seriously worried about factoring machines.

The harvest-now threat changes that calculus entirely. It converts QKD from a fascinating theoretical tool into a practical necessity for anyone managing secrets with long lifespans. The physics that makes QKD work- the indescribable fragility of quantum states in the face of measurement- turns out to be exactly the right property for defeating an adversary who plans to collect first and decrypt later.

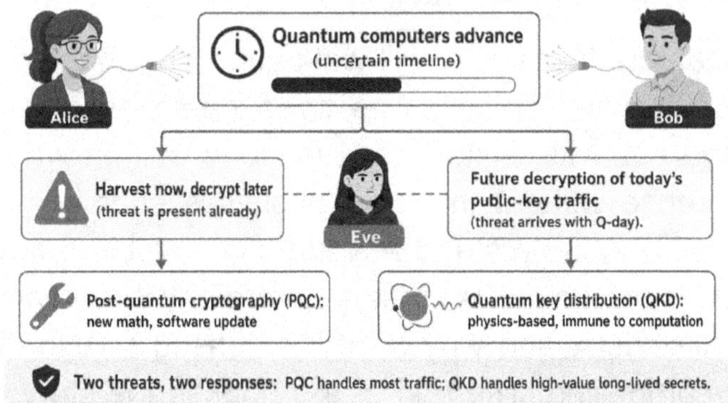

Diagram 2.12: The threat timeline and the response landscape

The two responses are not in competition. Post-quantum cryptography is a software upgrade that can be deployed across the entire internet through standard update channels. It is the right answer for the overwhelming bulk of everyday encrypted traffic. QKD is a hardware-based technology that requires physical infrastructure, fiber-optic links or satellite connections, and is suited to the highest value. In these longest-lived communications, even the theoretical risk of a future mathematical breakthrough is unacceptable. Together, they provide defense-in-depth against the quantum threat.

Understanding both options, and when to reach for each, is what separates an informed decision-maker from one who either panics or dismisses the risk. This book focuses primarily on QKD because its physics are fascinating and its role is often misunderstood. Still, every chapter that touches on deployment will keep one eye on how QKD and PQC work together rather than treating them as rivals.

The chapters that follow will walk you through exactly how QKD works, how it has been deployed in real networks, and its limitations. But none of that would make sense without the foundation this chapter has built. You now understand why there is a problem, who is at risk, why the risk is present before any threatening machine exists, and why a physics-based solution is newly compelling. The rest of the book is the story of that solution.

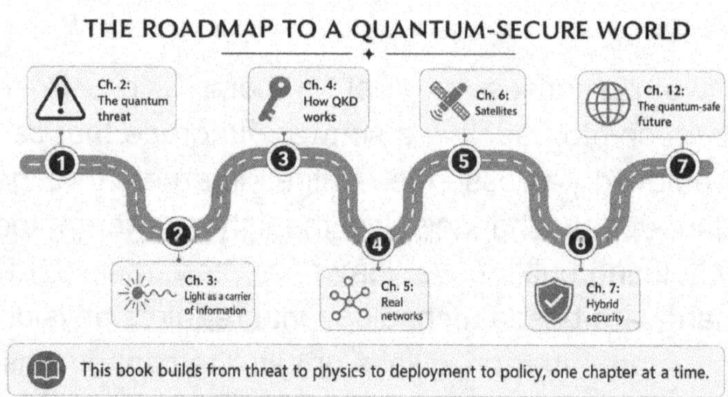

Diagram 2.13: The road ahead in this book

Key Concepts Checklist

• I can explain why a quantum computer is different from an ordinary computer, using the idea of superposition and the maze analogy.

• I understand that Shor's algorithm specifically threatens public-key encryption, not all encryption.

• I know that symmetric encryption with adequately long keys is far more resilient against quantum attacks.

• I can explain "harvest now, decrypt later" to someone who has never heard of it, including why it makes the threat present even before quantum computers arrive.

• I understand how to estimate a secret's urgency by combining its required lifetime with the quantum arrival window and the time needed to migrate.

• I can say why uncertainty about the timing of quantum computers argues for acting early rather than waiting for confirmation.

• I understand why QKD is a complement to post-quantum cryptography and what gap it fills.

Questions to Explore

• What secrets in your own life or organization would still be damaging if exposed in fifteen years? Which ones are currently protected only by public-key methods?

• Why might "we don't know exactly when" be more alarming than being told a fixed deadline? How does uncertainty change how an organization should plan?

• Should organizations act on a threat that has not fully arrived? What are the practical costs of waiting compared with the costs of acting early, and how would you make that case to a budget committee?

• How would you explain the harvest now threat to a colleague who dismisses quantum risk as science fiction? What single fact or analogy would you reach for first?

Try This / Reflect

Make a short personal or organizational inventory. Write down three categories of information you or your organization would want kept secret for more than ten years. For each one, note where that information is stored, who might have an interest in obtaining it, and whether encrypted copies of it could be accumulating somewhere right now. Then ask: Is the system protecting that information using public-key cryptography? If so, your shelf life calculation applies. This exercise takes fifteen minutes and makes the abstract risk feel very concrete very quickly.

If you want to go further, look up whether the software systems you rely on most have published a post-quantum migration roadmap. Many major technology vendors have done so or are in the process of doing so. Seeing a real roadmap from a real provider often does more to establish the credibility of the threat than any explanation.

Glossary Terms

• **Qubit -** The basic unit of information in a quantum computer, capable of existing in a superposition of zero and one before being measured, unlike a classical bit, which is always one or the other.

• **Shor's algorithm -** A procedure for quantum computers that can factor large numbers and solve the discrete logarithm problem far faster than any known classical method, making it capable of breaking public-key encryption on a sufficiently powerful machine.

• **Harvest now, decrypt later -** A strategy in which an adversary records encrypted data today, before they can decode it, and stores it until a quantum computer powerful enough to decrypt it becomes available.

• **Public-key cryptography -** A family of encryption methods in which two mathematically related keys (one public, one private) are used to exchange information securely; the security relies on certain math problems being computationally hard.

• **Shelf life of a secret -** The length of time a piece of information must remain confidential; when combined with the quantum arrival window and migration time, it determines how urgently a given system needs quantum-safe protection.

• **Post-quantum cryptography (PQC) -** A field of cryptography focused on developing mathematical algorithms that are believed to be resistant to attacks by quantum computers, designed to replace current public-key standards.

Key Takeaways

• A large enough quantum computer running Shor's algorithm could break the public-key encryption protecting most internet traffic, including HTTPS connections, email key exchanges, and financial transactions.

• Symmetric encryption is far more resilient; using longer keys is sufficient defense, so the threat is targeted at key exchange, not at the symmetric locks themselves.

• Harvest now, decrypt later means that long-lived secrets, health records, defense information, intellectual property, and diplomatic communications are already at risk even though no cryptographically relevant quantum computer currently exists.

• No one can name the exact arrival date of a threatening quantum computer, and that uncertainty argues for migrating the most sensitive systems early rather than waiting for confirmation.

• Quantum key distribution offers a physics-based alternative to math-based key exchange, one that is immune to computational attacks of any kind, making it uniquely suited to defeating the harvest-now strategy for high-value, long-lived secrets.

3 A Gentle Introduction to the Quantum World

Overview

Quantum key distribution (QKD) rests on a small set of rules about how very small things behave. These rules feel strange at first because nothing in ordinary life prepares us for them. But the engineers who build quantum networks treat these rules as reliable tools, the way a carpenter trusts the straightness of a level or the sharpness of a blade. The goal of this chapter is to translate those rules into plain language so that you, a reader with no physics background, can walk away with exactly the toolkit you need. The concepts covered here are the photon, polarization, superposition, measurement disturbance, the uncertainty principle, and the no-cloning theorem. Each one is introduced through an everyday analogy, and each analogy is honest about where it stretches thin.

This chapter is also careful about what it leaves out. Popular science articles often portray quantum physics as a swirl of mysticism, parallel universes, and observers who conjure reality into existence. None of that belongs in a practical guide to QKD. The ideas presented here are engineering facts. They are the reason an eavesdropper cannot intercept a quantum key without detection, and they are the reason the protocol you will meet in the next chapter actually works. By the time you finish reading, you will not be confused or overwhelmed. You will have a small, solid, and complete foundation, and you will be ready for everything that follows.

Opening Scenario

Imagine you are watching a nature documentary about magnetism. The narrator explains that every compass needle in the world points north because the Earth itself acts like a giant magnet buried beneath your feet. You cannot see that invisible magnetic force, but you can trust it, because every compass you have ever seen confirms it. Now imagine a second documentary, this one about quantum physics. The narrator says that particles can be in two places at once, that the act of looking at something changes it, and that nothing is real until it is observed. You watch, slightly stunned, and wonder whether physicists have lost their minds.

The confusion comes from a collision of language. The words "observation," "reality," and "two places at once" carry enormous everyday baggage. When physicists use them, they mean very specific, narrow, technical things. When documentary writers use them, they mean everything at once, and the drama that follows is real but misleading. The engineers building actual quantum networks do not spend their days pondering whether reality exists. They spend their days measuring pulses of light and counting errors. The quantum effects they rely on are real, repeatable, and predictable. What this chapter does is strip away the drama and hand you only the useful facts.

Think of this chapter as a short vocabulary lesson before a foreign film. You do not need to speak the whole language fluently. You need about six words, understood deeply and correctly, and the film will make complete sense. Those six words are: photon, polarization, superposition, measurement disturbance, uncertainty, and no-cloning. Learn them here, and the rest of this book opens up like a clear road on a sunny morning.

What This Chapter Will Cover

• Light as Particles: The Photon. The photon is the smallest possible unit of light, a single indivisible packet. The fact that a photon cannot be split in half is central to how QKD protects its keys.

• Polarization as a Direction. A photon's polarization is its orientation, like the direction an arrow points. QKD encodes the secret key bits by choosing which direction each photon points before sending it.

• Superposition Without the Hype. Superposition is a genuine blend of possibilities before measurement. The spinning-coin analogy introduces the idea clearly, with an honest note about where the analogy stops being useful.

• Measurement Changes Things. Measuring a quantum object forces it to settle on one outcome and disturbs it in the process. This single rule is the bedrock of QKD's eavesdropping detection.

• The Uncertainty Principle, Plainly. Some pairs of properties cannot both be known simultaneously. This prevents an eavesdropper from quietly learning everything about a photon before passing it along.

• The No-Cloning Rule. An unknown quantum state cannot be perfectly copied. This closes the loophole an eavesdropper might hope to exploit by copying the key in transit.

• What You Now Have and What You Don't Need. A summary of the small set of quantum facts you now own, with reassurance that this is the complete toolkit for understanding QKD.

Light as Particles: The Photon

For most of human history, scientists argued about whether light was a wave or a stream of particles. In everyday life, light certainly behaves like a wave: it bends around corners, it ripples through water, and it creates the rainbow patterns you see on soap bubbles. So the wave picture held sway for a long time, and it explained a great many things beautifully. Then, about a century ago, a series of careful experiments revealed something unexpected. When scientists turned the brightness of a light source down far enough, the light did not fade smoothly into darkness the way a wave would. Instead, it arrived in tiny, individual clicks, like raindrops tapping on a tin roof rather than a steady drizzle. Each click was a single, indivisible packet of light. Physicists called it the photon.

You can picture a photon as the world's smallest coin of light. Just as a coin cannot be cut into a half-coin that still spends as money, a photon cannot be divided into a half-photon that still behaves as light. It is the atom of brightness, the minimum possible amount. A candle flame sends out an unimaginable number of them every second, billions upon billions, which is why candlelight looks smooth and continuous. But the smoothness is the same illusion you see in a cinema, where a rapid sequence of still images appears to be a flowing movie. Look closely enough, and discrete units are there.

This indivisibility turns out to be enormously important for security. Consider what happens when a thief tries to tap into a beam of classical light carrying a message. The beam contains so many photons that the thief can peel off a tiny fraction, read that fraction, and the sender and receiver never notice the loss. The beam arriving at the destination is almost unchanged. This is roughly how a fiber-optic wiretap works in the classical world. Now imagine the security system sends only a single photon to carry each bit of the key. There is nothing to peel off. You either take the whole photon, in which case the intended receiver gets nothing and immediately knows something went wrong, or you leave it alone. The indivisibility of the photon is not an accident of design; it is the first wall in the security architecture.

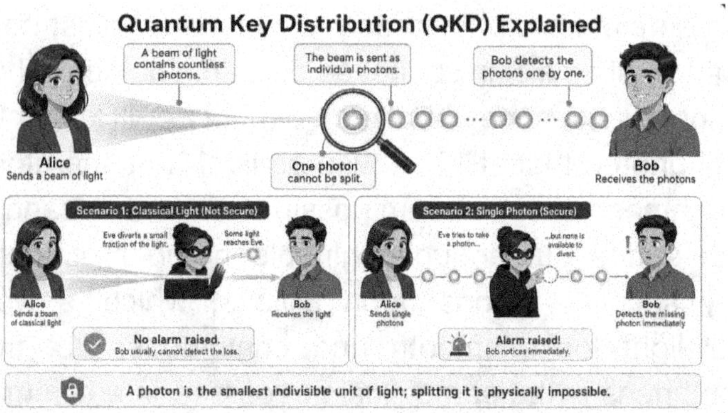

Diagram 3.1: The photon as the smallest coin of light

The practical upshot is that quantum communication channels are naturally stingy. They send one photon at a time, and each one is precious. When Alice, the sender, fires a photon at Bob, the receiver, that photon is all there is. The channel carries no spare copies, no backup signal, no redundant information to be harvested. This stinginess is not a limitation to be engineered around. It is the feature that makes the whole enterprise of quantum key distribution (QKD) possible in the first place.

There is one nuance worth naming early. Real-world QKD systems are not yet perfect at sending exactly one photon per pulse. Sometimes a pulse contains two photons by accident. Security researchers have studied this carefully, and there are protocols designed to handle it. But the ideal we are building toward, and the ideal that captures the security logic, is the single-photon channel. Hold that picture in mind as you read the rest of this chapter.

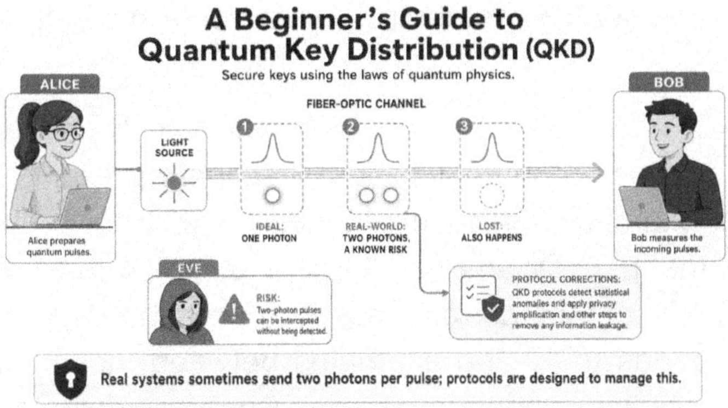

Diagram 3.2: Single-photon versus multi-photon pulses

Polarization as a Direction

Every photon carries a property called polarization. Polarization is, at its simplest, an orientation in space. Think of it as the direction an invisible arrow points when it is attached to the photon. That arrow can point straight up, or tilted to the right at 45 degrees, or horizontal like a horizon line, or any other angle around the clock. The photon itself is always the same particle; it is the orientation of the arrow that changes.

A good everyday model is the slat of a window blind. You can tilt the slats to let in light from a narrow band of angles and block everything else. If you turn the slat so it is perfectly vertical, it passes light that is vibrating vertically but blocks light vibrating horizontally. Rotate the slat to 45 degrees, and it passes light vibrating along that diagonal. This is exactly how a polarizing filter works in sunglasses or a camera lens, and it is exactly how a polarizing detector works in a QKD system. The filter is the measuring device, and the angle you set it to determines what you are willing to accept.

In QKD, Alice sends photons with deliberately chosen polarizations. Each polarization direction represents a bit: for example, vertical might mean 0, and horizontal might mean 1. Or she might use a diagonal basis in which 45 degrees corresponds to 0 and 135 degrees corresponds to 1. She alternates between these two basis sets in a random order, and Bob uses detectors that are also set to one basis or the other. The matching and mismatching of bases is the heart of the protocol you will meet in the next chapter. For now, the key point is simple: polarization is a direction associated with a photon; Alice can choose it, and Bob can measure it.

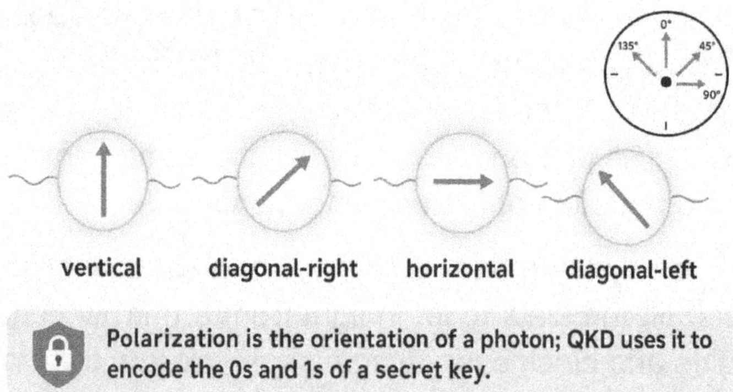

Diagram 3.3: Polarization as an arrow on a photon

The beauty of polarization as a carrier of information is its precision. Alice can set a photon's polarization to within a fraction of a degree. The direction is a real, physical property that Bob's detector can measure. And crucially, as you will see in the sections on superposition and measurement, an eavesdropper who intercepts the photon and tries to read that direction will disturb it in ways that cannot be hidden. The direction is readable, but it is not secretly readable. That asymmetry is what makes the whole system work.

You might wonder how sensitive a real detector is. Modern single-photon detectors can distinguish vertical from horizontal polarization with very high accuracy. They are not perfect, and the small error rate they introduce is measured and accounted for. But the principle is clean: a correctly oriented detector clicks for a matching polarization and stays silent for a mismatched one, just as a vertical blind slat passes vertical light and blocks horizontal light.

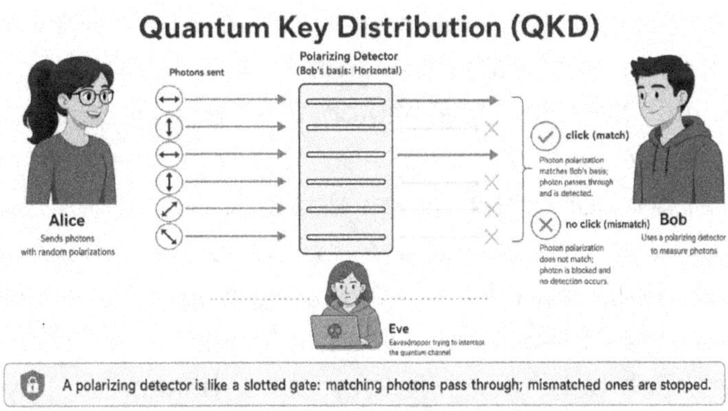

Diagram 3.4: Polarizing filter passing and blocking photons

Superposition Without the Hype

Here is the quantum idea that receives the most dramatic treatment in popular science and is also the most distorted. Superposition is real, useful, and not magic. Let us start with what it actually is.

Imagine you spin a coin on a table. While it is spinning, it is neither heads nor tails. It is in a genuine sense both, or more accurately, it is in a state that has no single definite answer. When the coin slows and falls flat, it becomes definitely heads or definitely tails. Before it falls, asking "is it heads?" has no clean answer, because the coin has not yet committed. Physicists call that uncommitted, spinning state a superposition: a blend of possibilities that has not yet resolved into one outcome.

A quantum particle like a photon can be in something like that spinning state with respect to a measurement. If Alice prepares a photon with a 45-degree diagonal polarization and Bob measures it with a vertical-or-horizontal detector, the photon is in a superposition from that detector's point of view. It is genuinely neither vertical nor horizontal. When Bob's detector fires, the photon randomly picks one or the other, and any hidden information does not determine which choice it makes. It is a fresh, irreducible randomness baked into nature. Bob gets a click for vertical or horizontal, fifty-fifty, with no prior clue as to which it will be.

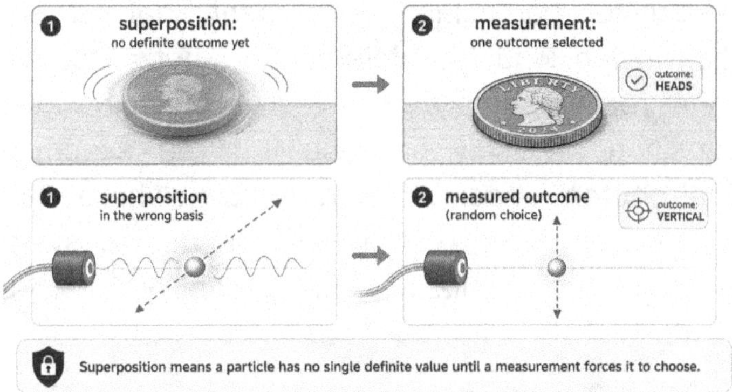

Diagram 3.5: The spinning coin as a superposition

Now for the honest note about where the spinning-coin analogy breaks down, because this matters. A spinning coin is still a physical object obeying classical physics. In principle, if you measured its exact tilt and spin at every moment, you could predict whether it would land heads or tails. The outcome is unknown to you but determined by the physics. A photon in superposition is different. Its outcome is not just unknown to you; it is genuinely undetermined even in principle. No hidden measurement could have predicted it. That is a distinction the analogy cannot capture, but it does not hurt the practical picture. For QKD, what matters is the outcome: when measured in the wrong basis, the photon's answer is random. Random outcomes cannot be predicted, and that unpredictability, whether classical or quantum, is what the security relies on.

A second place where the analogy misleads is the idea of "two things at once." The photon in superposition is not secretly in two places or secretly pointing in two directions. It is in a state that has no definite direction with respect to a particular measurement. That is a very specific, limited kind of indeterminacy, not a cosmic claim that reality is fuzzy everywhere. Engineers love superposition precisely because it is so exact. A diagonal photon measured by a vertical-or-horizontal detector will always give a fifty-fifty random answer, every single time, reliably. That reliability is what makes it useful.

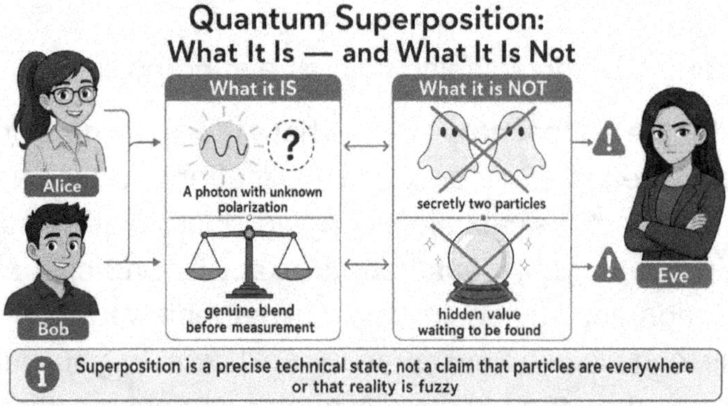

Diagram 3.6: What superposition is and is not

Measurement Changes Things

Of all the quantum ideas in this chapter, this one has the most direct security consequence, so it deserves the most careful attention. In everyday life, measuring something does not have to disturb it. You can read the temperature of a warm cup of coffee by glancing at a thermometer, and the act of reading changes the temperature by such a tiny amount that no one cares. You can weigh a bag of flour without changing how much flour it contains. You can photograph a landscape without moving any of the mountains.

Measurement in the classical world is, for most practical purposes, passive observation.

Quantum measurement is fundamentally different. When you measure a quantum object, you are not passively recording a value that was already there. You are interacting with the object in a way that forces it to take on a definite value, and that interaction changes the object. The spinning coin, once it falls flat, cannot be made to spin again just by reading which side it landed on. Once the photon resolves its polarization in response to a detector, that particular act of resolution is irreversible. The photon that comes out of the measurement is not the same as the photon that went in.

Picture a soap bubble. The bubble has a surface with a swirling, shifting pattern of colors. Now imagine you try to measure the exact color of a single spot by pressing your fingertip against the bubble. The act of touching pops the bubble, or at minimum warps it so profoundly that the color pattern you see afterward is not the one you were trying to read. The measurement is not gentle. It is an intervention. This is a rough but useful image for what happens when a detector intercepts a photon: the photon's state is changed in a way that cannot be undone or hidden.

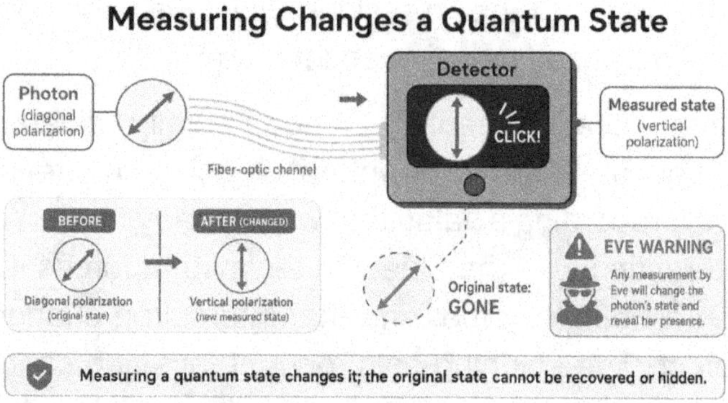

Diagram 3.7: Measurement as an irreversible intervention

Now connect this to eavesdropping. Call the eavesdropper Eve. Eve is sitting somewhere along the fiber between Alice and Bob, and she intercepts one of Alice's photons. She wants to learn the photon's polarization and then send an identical photon onward to Bob so he never suspects anything. The problem is the measurement step. When Eve measures the photon, she must choose a detector basis. If she chooses the wrong one, the photon is now in a new state determined by her measurement, not Alice's original preparation. Eve cannot measure it without changing it. She cannot read the original polarization and then faithfully reproduce it, because the measurement itself replaced the original state with a new one. When that changed photon reaches Bob, it will produce errors that Alice and Bob can detect by comparing a sample of their data.

This is not a technical limitation of Eve's equipment that better technology might fix. It is a consequence of the physical law that measurement changes quantum states, a law that applies to every possible measuring device in the universe, no matter how sophisticated. Eve is not being foiled by clever engineering. She is being foiled by physics. And that is what makes QKD different from every classical security system: the physical rules of the universe are doing the work, not the difficulty of a math problem.

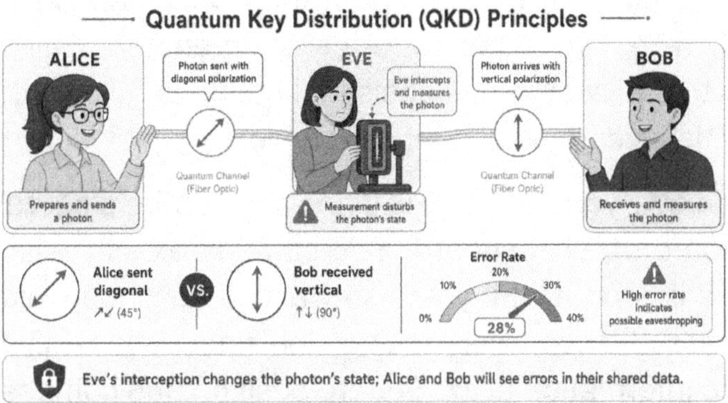

Diagram 3.8: Eve's measurement disturbs the photon

The Uncertainty Principle, Plainly

The uncertainty principle is one of the most famous ideas in all of science, and also one of the most misunderstood. Its popular reputation suggests it is about the clumsiness of our instruments, that we cannot measure things precisely because our tools are too crude. The actual principle is subtler and more powerful. It says that certain pairs of properties are linked fundamentally: knowing one of them precisely forces the other to be genuinely indeterminate, not just unknown to us, but undetermined in nature itself.

The classic pairing is position and momentum, which is the combination you encounter in most physics courses. For QKD, the relevant pairing is between different polarization bases. Think of it this way: a photon prepared in the diagonal basis has a perfectly defined diagonal polarization, but its behavior in the vertical-or-horizontal basis is completely random. You can know everything about it in the diagonal world and know nothing useful about it in the horizontal-vertical world. The two bases are complementary. Nailing down the answer in one basis makes the answer in the other basis completely unpredictable.

A good everyday analogy is trying to describe a figure's position on a chessboard. You can say exactly which square it is on. But if you also want to describe which compass direction it is facing, the chessboard gives you no information at all. Square position and compass bearing are properties drawn from completely different systems of description. Knowing one says nothing about the other. In quantum mechanics, certain pairs of properties are like this, except the relationship is even tighter: measuring one actively destroys the information about the other, rather than merely failing to provide it.

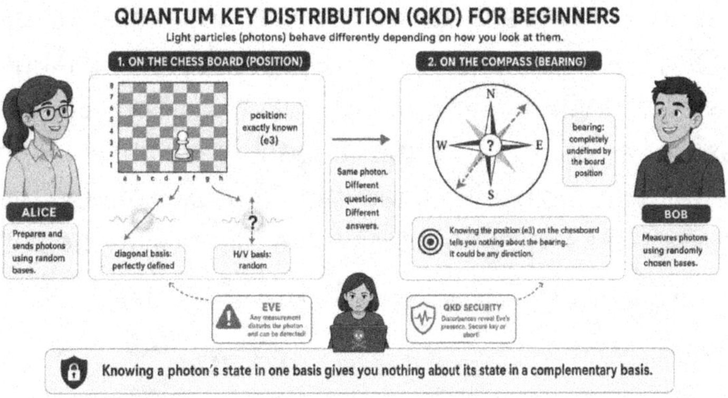

Diagram 3.9: Complementary bases like a chess board and a compass

For Eve, this creates an impossible problem. She does not know in advance which basis Alice used to prepare each photon. She has to guess. If she guesses wrong, she measures in the wrong basis, gets a random answer that tells her nothing about Alice's intended bit, and she changes the photon's state in the process. Even if she could somehow learn the correct basis for one photon, that knowledge would come too late to be useful for the next one, because Alice chooses each basis randomly. The uncertainty principle is not just a nuisance for Eve; it is an absolute barrier. She cannot gather full information about the photon without revealing herself through the errors she introduces.

It is worth pausing here to appreciate how elegant this is. The security of QKD is not built on a long password that a computer might one day crack. It is built on the geometry of measurement. The fact that two bases are complementary is a mathematical truth about the structure of quantum mechanics. It will not become easier to circumvent as computers get faster or as algorithms improve. A faster computer cannot change the geometry. A more powerful algorithm cannot measure a photon without choosing a basis. The uncertainty principle is baked into the fabric of how the universe works, and no future technology will repeal it.

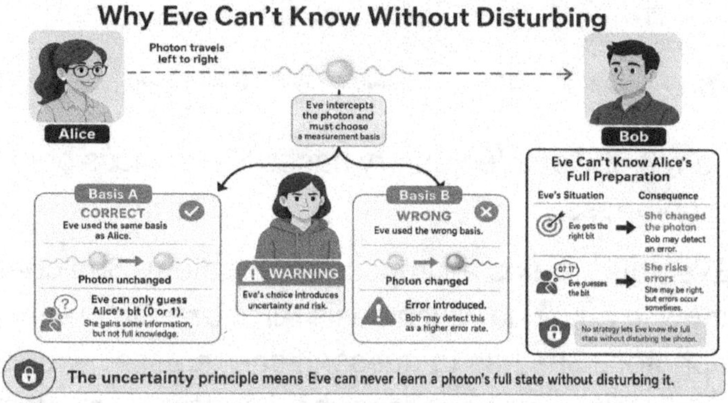

Diagram 3.10: Uncertainty bars Eve from gathering full information

The No-Cloning Rule

Recall from Chapter 1 the attack that seemed so tempting: instead of measuring the photon and risking the state change, what if Eve copied it? She could make a perfect duplicate of the photon, measure the copy quietly, and let the original travel on to Bob untouched. This seems like a clean workaround. It would sidestep measurement disturbance entirely. If it worked, QKD would be broken.

It does not work. In 1982, two physicists proved a theorem, now called the no-cloning theorem, which says it is physically impossible to make a perfect copy of an unknown quantum state. Not impractical. Not difficult. Impossible. The proof is surprisingly short, and it follows from the same basic rules of quantum mechanics that govern everything else in this chapter. The argument, stripped of mathematics, goes roughly like this: to copy a state, you need to know what the state is. But learning what it is requires measuring it. And measuring it changes it. You cannot measure first and then copy, because measuring has already changed what you were trying to copy. The only copy you can make is a copy of the changed state, not the original. You are always one step behind.

Think of it like trying to trace a drawing while it is still being drawn. You need to see the drawing to trace it, but looking at it counts as a kind of interaction that slightly smears the original lines. The tracer can never reproduce the drawing exactly, because the original is never fully visible in a stable, un-interacted-with form. The drawing is always in motion as long as you are looking.

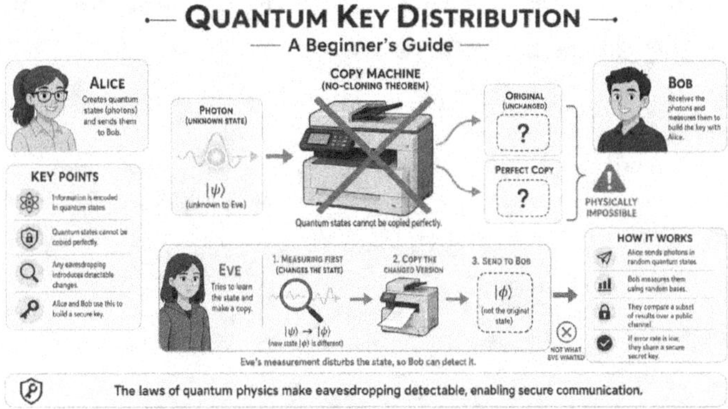

Diagram 3.11: The no-cloning trap

The no-cloning theorem closes the most obvious loophole in QKD security. Combined with measurement disturbance and the uncertainty principle, it forms a three-part wall. Eve cannot measure without disturbing. Eve cannot learn which basis to use without already having the photon. Eve cannot copy the photon and measure the copy later. Each loophole closes the others, and together they leave Eve with no path to a silent attack. She must interact with the photons, and every interaction leaves a mark.

There is one subtlety that is worth acknowledging. The no-cloning theorem applies to unknown states. If Alice always used the same polarization for every photon, Eve could eventually figure out the pattern by watching the error statistics. Then she would not need to clone: she would already know. This is exactly why Alice randomizes her basis choices for each photon. The randomness is not optional decoration. It is the ingredient that keeps each photon's state unknown to Eve, which is the condition the no-cloning theorem requires to do its protective work.

Diagram 3.12: Three walls Eve cannot climb

What You Now Have and What You Don't Need

You have now met every quantum concept that matters for understanding QKD. Let us do a quick inventory, because the list is shorter than you might expect.

A photon is the smallest indivisible particle of light. Its indivisibility means a single-photon channel has nothing to steal without the theft being noticed. Polarization is the direction a photon points, and it is the property Alice uses to encode each bit of the key. Superposition means a photon can be unresolved before measurement, and that unresolved state collapses into a random outcome when a detector fires in the wrong basis. Measurement disturbance means that the act of reading a quantum state changes it, irreversibly and detectably. The uncertainty principle means that knowing a photon's state in one basis tells you nothing about its state in the complementary basis.

And the no-cloning theorem means a perfect copy of an unknown quantum state is physically impossible.

That is the complete list. You do not need to understand entanglement for the main QKD protocol in this book, though it appears in a more advanced variant covered later. You do not need to know about wave functions, Hamiltonians, or Hilbert spaces. You do not need the Schrödinger equation or the Born rule. The six concepts above are necessary and sufficient. If you can explain them in your own words, you have everything needed to follow the full protocol in the next chapter and every chapter after it.

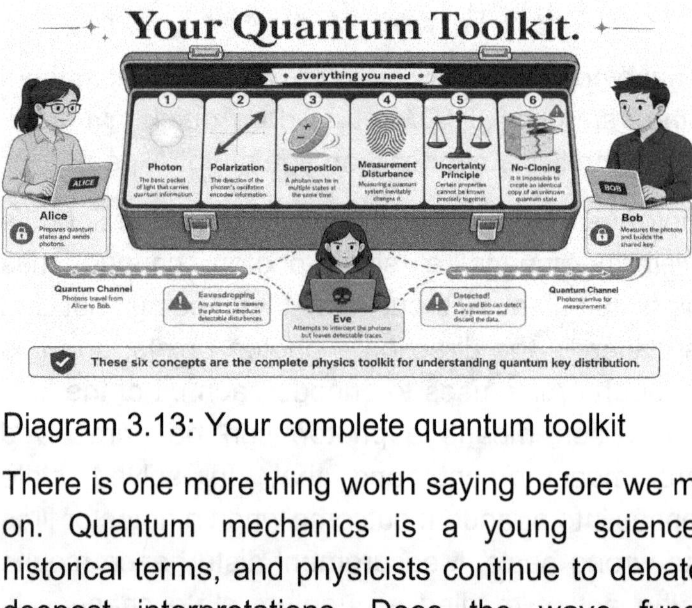

Diagram 3.13: Your complete quantum toolkit

There is one more thing worth saying before we move on. Quantum mechanics is a young science in historical terms, and physicists continue to debate its deepest interpretations. Does the wave function collapse? Are there many worlds? Is the randomness truly fundamental? These are fascinating questions, and smart people disagree about the answers. But here is the good news: none of those debates change the practical predictions of quantum mechanics. Every experiment ever performed confirms that measurement disturbs quantum states, that complementary bases are genuinely complementary, and that unknown states cannot be cloned. Whatever the philosophical truth turns out to be, the engineering rules are solid. You can build a working QKD system today without resolving a single philosophical question about the foundations of reality.

Think of it like driving a car. You do not need to know how the internal combustion cycle converts fuel into torque to drive safely and skillfully. The car works whether or not you understand it at the level of thermodynamics. QKD works whether or not you understand quantum mechanics at the level of interpretation theory. The rules you have learned in this chapter are the driving manual, not the mechanical engineering textbook. That is all you need, and it is exactly enough.

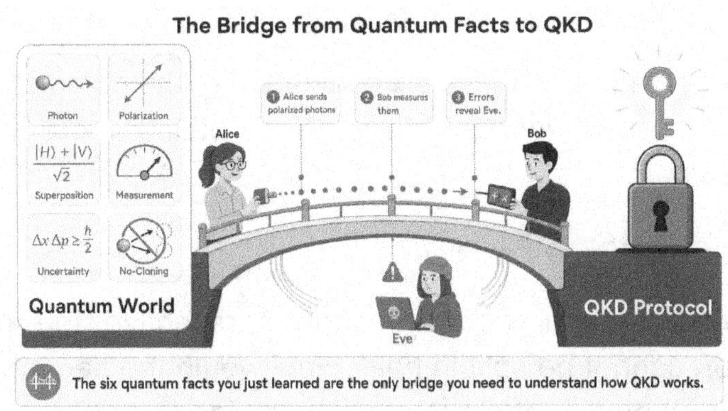

Diagram 3.14: The bridge from quantum facts to QKD protocol

Key Concepts Checklist

• I can describe what a photon is and explain why its indivisibility matters for communication security.

• I understand polarization as a direction attached to a photon, and I know that QKD encodes key bits using that direction.

• I can explain superposition using the spinning-coin analogy, and I can name at least one way the analogy breaks down.

• I understand that measuring a quantum object changes it in a way that cannot be hidden or reversed.

• I can state the uncertainty principle in one plain sentence: knowing a photon's state in one basis makes its state in the complementary basis completely random.

• I can explain why an unknown quantum state cannot be perfectly copied, and I can connect this to why Eve's copy-and-measure attack fails.

• I understand that these six concepts are the complete physics toolkit for QKD and that nothing more is required.

Questions to Explore

• Why is it useful, from a security perspective, that a photon cannot be split in half? How would the security picture change if photons could be divided?

• The spinning-coin analogy for superposition is imperfect. Where exactly do you think it breaks down, and does the breakdown matter for understanding QKD security?

• If measuring a quantum object always changes it, what does that imply for any secret surveillance of a quantum signal? Is there any conceivable way around this constraint?

• Why might the no-cloning theorem be considered a more fundamental protection than any encryption algorithm, however strong?

Try This / Reflect

Find two pairs of polarized sunglasses, or watch a short online video demonstrating polarized light (search for "crossed polarizers experiment"). Hold one lens in front of the other and slowly rotate it. Watch the view through the two lenses darken and brighten as you turn. When the lenses are aligned, light passes freely. When they are at right angles to each other, almost no light gets through. You are watching photon polarization with your own eyes. The property that dims and brightens the view is the same property Alice uses to encode the 0s and 1s of a secret key. Real QKD hardware uses exactly this principle, scaled down to one photon at a time.

If you want to go a step further, find a second pair of polarized sunglasses and try inserting one lens from one pair between the two crossed lenses at a 45-degree angle. The total darkness you had before should partly brighten again. This is a glimpse of the diagonal basis: a third orientation that interacts with both the vertical and horizontal ones in a way that a simple classical explanation cannot fully capture. You do not need to understand why it happens to appreciate that quantum polarization is a real, touchable phenomenon, not a textbook abstraction.

Glossary Terms

- **Photon -** The smallest indivisible unit of light; a single particle of light that cannot be divided into smaller pieces while still behaving as light.
- **Polarization -** The orientation, or direction, of a photon's oscillation; in QKD, this direction is used to encode the bits of a secret key.
- **Superposition -** A quantum state in which a particle has no single definite value for a particular property before measurement; when measured, the particle randomly resolves into one of the possible outcomes.
- **Measurement disturbance -** The rule that observing a quantum object forces it to take on a definite value and changes it in the process; this change cannot be hidden or reversed.
- **Uncertainty principle -** The rule that certain pairs of quantum properties, such as polarization in two complementary bases, cannot both be known precisely at the same time; learning one destroys information about the other.
- **No-cloning theorem -** The physical law stating that it is impossible to make a perfect copy of an unknown

quantum state; this prevents an eavesdropper from copying a photon in transit without detection.

Key Takeaways

• QKD relies on only six quantum facts, all understandable through everyday analogy and all supported by decades of experimental confirmation.

• A photon is the indivisible atom of light; its indivisibility means a single-photon channel has nothing to steal without the theft being noticed.

• Measuring a quantum object changes it in an unavoidable and detectable way, making silent eavesdropping physically impossible.

• The uncertainty principle prevents an eavesdropper from learning a photon's full state, because knowing its value in one basis destroys information about the complementary basis.

• The no-cloning theorem closes the copy-and-measure loophole, ensuring that the only way to gain information about a photon is to interact with it, and every interaction leaves a mark.

4 How Quantum Key Distribution Actually Works

Overview

This chapter is the payoff for everything that came before it. The quantum rules you met in Chapter 3 (no cloning, measurement collapses the state, and polarization as a real physical property of light) are not abstract curiosities. They are the machinery that makes quantum key distribution (QKD) possible. Here you will watch those rules in action as we walk through BB84, the original and still the most widely taught QKD protocol, from the first photon to the finished key. Two characters, Alice and Bob, will guide you through every step. By the time you finish this chapter, you will be able to describe the entire process in your own words, and you will understand why the result is genuinely secure.

The central insight of this chapter is something worth stating right at the top: the secret key is never sent. Alice never packages up a string of secret bits and ships it to Bob. Instead, the key emerges from a series of random choices Alice and Bob make independently, combined with the strange behavior of quantum particles. The key is built, not delivered. That shift in thinking is the heart of BB84, and it is why the protocol solves a problem that stumped cryptographers for decades. Everything else in this chapter is an explanation of how that building process works, step by careful step.

Opening Scenario

Imagine two colleagues, Alice and Bob, who work for a hospital system. Alice is in the main campus downtown, and Bob is at a satellite clinic across the city. They need to share patient records securely over a network. Ordinary encryption would work, but they are worried about the future: a powerful enough computer, perhaps one that does not yet exist, could one day break the keys they are using today if someone has been quietly recording their traffic. They need a key-sharing method whose security does not depend on assumptions about computing power. They need something grounded in physics.

Their IT team has just installed a fiber link between the two buildings and connected it to a small device capable of sending and detecting single photons. That fiber is the quantum channel. They also have their ordinary phone and internet connection, the classical channel. Alice sits down at her terminal and begins sending photons. She is not sending the key. She is not even sending data in any traditional sense. She is sending a long stream of individually polarized particles of light, each one the result of a coin flip she just made. Bob, at the other end, is measuring each photon with a detector that also makes a coin flip of its own. Neither of them knows what the other is choosing. And yet, by the time they are done and have had a short public conversation, they will share a secret string of bits that no one else on the planet can reconstruct.

That is the promise of BB84. It sounds almost magical. The rest of this chapter shows you exactly why it works, one step at a time, without any mathematics, using nothing more than the physical intuitions you already have about coins, filters, and light.

What This Chapter Will Cover

• **Meet Alice and Bob:** The two standard characters of cryptography, their goal, and the simple two-channel setup that every BB84 exchange requires.

• **The Two Channels QKD Needs:** Why one channel carries photons while the other carries ordinary conversation, and why the public channel cannot leak the secret.

• **Step One: Alice Sends Random Photons:** How Alice chooses a random bit and a random basis for every photon she sends, and what that stream looks like from a bird's-eye view.

• **Step Two: Bob Measures Randomly:** How Bob picks his own random basis for each measurement, what happens when his choice matches Alice's, and what happens when it does not.

• **Step Three: Sifting the Key:** The short public conversation where Alice and Bob reveal only their bases (never their bits), and the rule for deciding which photons to keep.

• **Step Four: Checking and Cleaning:** How Alice and Bob use a sample of their shared bits to detect eavesdropping, then correct residual errors and distill a clean final key.

• **Why the Key Is Built, Not Sent:** The big-picture view: why a key that was never in transit cannot be stolen in transit, and how the no-cloning rule and measurement rules from Chapter 3 protect every step.

• **The Full Picture at Once:** Putting all four steps together into a single coherent story, with a feel for timing and real-world scale.

Meet Alice and Bob

Almost every textbook on cryptography and quantum communication introduces two characters named Alice and Bob. The names are not accidental. They come from a 1978 paper that used A, B, and C as stand-ins for the parties in a message exchange, and the community quickly turned those initials into people. Alice and Bob have appeared in thousands of papers and textbooks since. They are the field's shorthand for "sender" and "receiver," and using them here avoids awkward phrasing such as "the first party" and "the second party."

Alice is the sender. In a BB84 exchange, she controls the device that generates and sends individual photons. She has a source of single photons and a way to orient each photon before sending it. Bob is the receiver. He controls the detector at the other end of the fiber. He measures incoming photons and records the results. Neither of them starts the process knowing anything secret. They do not share a password. They do not have a secret code book. They begin with nothing in common except their two-channel link and a shared agreement to follow the BB84 procedure.

That procedure was invented in 1984 by Charles Bennett and Gilles Brassard, and their initials plus the year give the protocol its name: BB84. It was the first protocol to demonstrate that quantum mechanics could be used as a practical key-distribution method. Before BB84, distributing a cryptographic key securely required either a trusted physical courier or a clever mathematical trick (and every mathematical trick carries some risk that a future computer will crack it). BB84 offered a third option: use the laws of physics themselves as the security guarantee.

The goal for Alice and Bob is simple to state. They want to end up with identical copies of a string of random bits, a string that nobody else knows. That string will become a key. They can then use that key to encrypt their patient records or any other sensitive communication. If they can reliably produce such a string, the security problem is solved. BB84 is the machine that produces it.

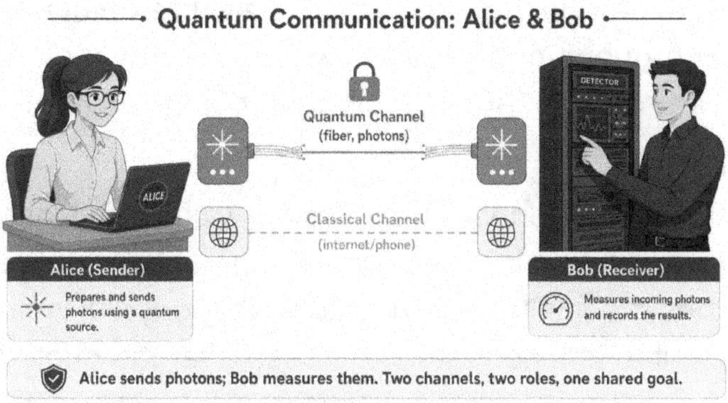

Diagram 4.1: Alice and Bob: the two-character cast

The Two Channels QKD Needs

One of the most common early questions about QKD is: if Alice and Bob are communicating openly over a classical channel, why cannot an eavesdropper listen to that conversation and figure out the key? The answer involves understanding what travels on each channel and what information is kept off both channels entirely.

The quantum channel carries photons. In practice, this is almost always an optical fiber, though some systems use the open air or even space. Each photon is a single particle of light, and it carries Alice's quantum information in the form of its polarization (the direction in which the light wave oscillates). The quantum channel is not a broadcast medium. A photon sent down the fiber either arrives at Bob's detector or it does not. If an eavesdropper tries to intercept and copy a photon, the no-cloning rule from Chapter 3 says this is impossible. If they try to measure the photon and then resend it, they disturb the photon's state in a detectable way. The quantum channel is where the raw material of the key travels, but the raw material is not the key itself.

The classical channel is an ordinary communication link. It could be an internet connection, a phone call, or any other channel you already use every day. Alice and Bob use it to talk about their process after the photons have been sent. Critically, anyone can listen to this conversation. The classical channel is assumed to be entirely public. The reason this is safe will become clear when you see exactly what Alice and Bob say on it: they announce which measurement tools (bases) they used, but they never announce the actual results (bits). The bases are the frame; the bits are the picture inside the frame. Giving away the frame reveals nothing about the picture.

Why does QKD need both channels? The quantum channel alone is not enough, because Alice and Bob need to compare notes to determine which photons to keep. The classical channel alone is not enough because ordinary channels offer no protection against eavesdropping. The two channels together create a system in which the valuable information (the bits) never travels on either channel. It sits quietly in Alice's and Bob's measurement records, waiting for the sifting conversation to reveal which records they share.

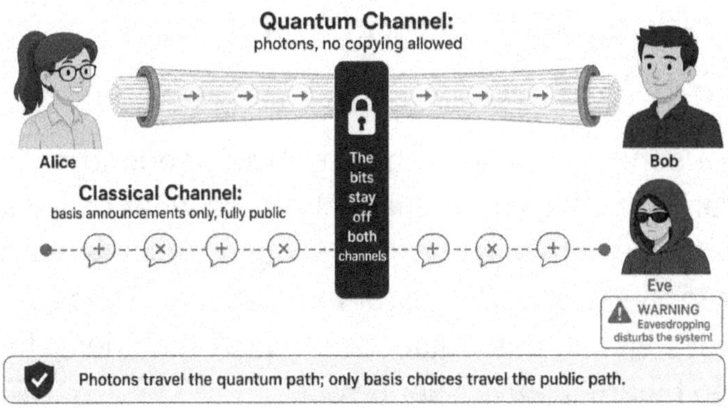

Diagram 4.2: Two channels, two jobs

There is a third piece of infrastructure worth mentioning: the assumption that Alice and Bob can verify each other's identity on the classical channel. If an adversary could impersonate Bob on the classical channel, they could interfere with the sifting conversation. In practice, Alice and Bob use a small amount of pre-shared information (a tiny authentication key) to verify identities on the classical channel. This is a much smaller problem than distributing a full encryption key, and it only needs to happen once at the start of the relationship. After that, each BB84 session produces new key material that can be used to authenticate the next session. Chapter 10 will return to this authentication requirement and explain why it represents one of QKD's most important real-world constraints.

Step One: Alice Sends Random Photons

Before sending a single photon, Alice needs to make two random choices. The first choice is what bit value to encode: a zero or a one. The second choice is which basis to use: the rectilinear or the diagonal. These two choices together determine exactly how she orients the photon before it leaves her device.

To understand what a basis is, picture a pair of sunglasses. Polarized sunglasses have lenses that only let through light oscillating in one specific direction. If you rotate the lens 90 degrees, it blocks light that the first orientation passed and passes light that the first orientation blocked. A "basis" in BB84 is exactly this kind of filter setup: a pair of orientations that are 90 degrees apart. Alice has two such pairs available.

The rectilinear basis uses vertical and horizontal polarization. A photon polarized straight up and down encodes one bit value (say, 1). A photon polarized straight left and right encodes the other bit value (0). The diagonal basis uses 45-degree-angle polarizations. A photon tilted 45 degrees to the right encodes 1, and a photon tilted 45 degrees to the left encodes 0. Each basis is a different "language" for expressing the same two-bit values. When both Alice and Bob use the same basis, they are speaking the same language. When they use different bases, communication gets scrambled.

So before sending each photon, Alice picks a bit at random (a coin flip: 0 or 1) and then picks a basis at random (another coin flip: rectilinear or diagonal). She orients the photon accordingly and sends it into the fiber. She records her two choices (bit and basis) for that photon in a private log. She does this for every photon in the stream, perhaps a thousand photons or a million, depending on how long a key she needs. Each photon is an independent event. No photon carries any information about the ones before or after it.

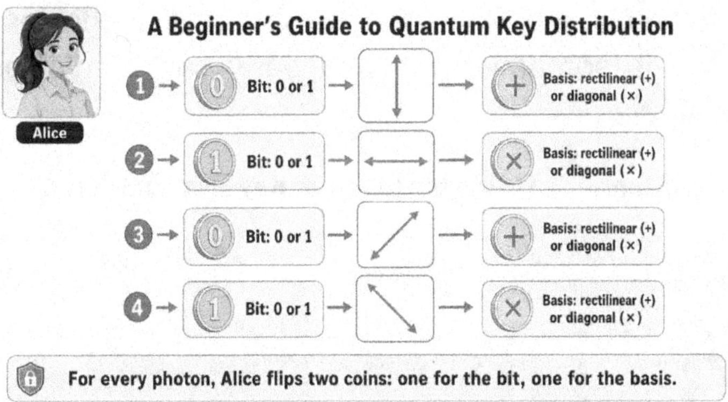

Diagram 4.3: Alice's two choices per photon

Here is an important point: Alice's log is strictly private. She is not encrypting it or hiding it with any clever trick. It simply never leaves her machine. She sends photons into the fiber, and the photons carry no information about their bit value to anyone who does not already know which basis to measure them in. Even if an eavesdropper intercepts every single photon, they face a problem: they do not know which basis Alice used for each one. We will come back to why that problem is insurmountable for the eavesdropper, but for now, follow Alice as she finishes her transmission and sits back to wait for the next step.

The stream of photons Alice sends is completely random from the outside. There is no pattern to the polarization angles, no repetition, no structure that a listener could use to reconstruct the original sequence of bit choices. This randomness is not a weakness; it is the design. The key's security ultimately rests on the fact that neither Alice nor Bob chose the key in advance. The key will emerge from the random intersection of their two independent sets of choices, and that is something no third party can predict or reconstruct.

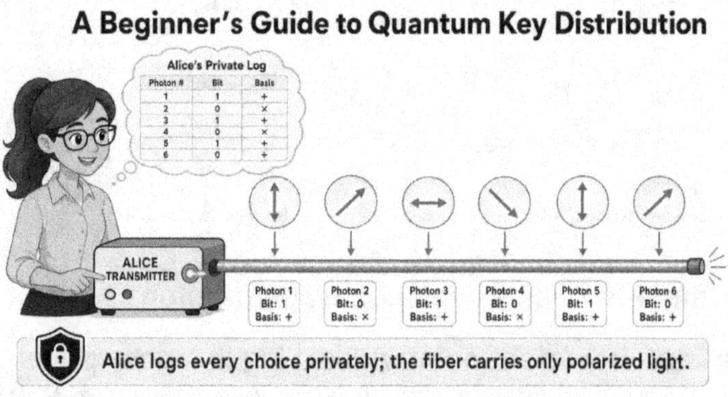

Diagram 4.4: Alice's stream of photons leaving the device

Step Two: Bob Measures Randomly

Bob's detector is running while Alice sends. For every photon that arrives, Bob must make a choice: which basis will he use to measure it? He has no way of knowing which basis Alice used for that photon. He has not communicated with Alice since the process began. So he does the only sensible thing: he picks a basis at random for each photon, just as Alice did, and records both his choice and the result he gets.

This is the pivotal moment in understanding BB84. When Bob's randomly chosen basis happens to match Alice's randomly chosen basis for the same photon, a remarkable thing happens: Bob's measurement result is guaranteed to agree with the bit Alice encoded. If Alice sent a vertical photon (encoding 1 in the rectilinear basis) and Bob measures in the rectilinear basis, he will always get a 1. The physics guarantees it. A photon measured in its own basis always yields the correct answer.

When Bob's basis does not match Alice's, the result is completely unpredictable. Imagine Alice sends a vertical photon (aligned with the rectilinear basis), and Bob tries to measure it in the diagonal basis. The detector has no rectilinear "slot" to recognize the photon's orientation. Instead, from the diagonal detector's point of view, a vertical photon is equally likely to register as either a 0 or a 1. The result is essentially a coin flip, totally random, completely unrelated to what Alice sent. Bob gets a number, but it carries no information about Alice's original bit.

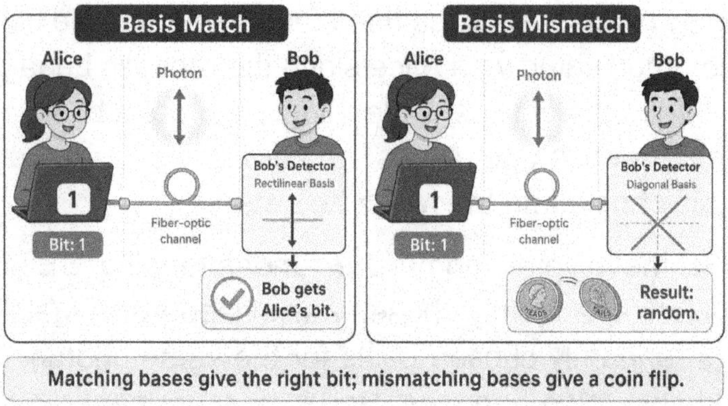

Diagram 4.5: Bob's measurement: match vs. mismatch

Bob's private log now holds two columns: his basis choice for each photon, and the bit value his detector registered. He does not yet know which of his measurements are reliable (the ones where he matched Alice) and which are noise (the ones where he did not). He has a long list of results. At this point neither Alice nor Bob has spoken. No information has traveled on the classical channel. The key-building process is entirely silent so far.

You might be wondering how often Alice's and Bob's bases will match. Because they each choose randomly and independently from the same two-basis menu, they will agree by chance about half the time. On average, roughly one in two photons will fall in the "match" category. This is not a problem; it is expected. The sifting step (coming next) will throw away the mismatches and keep the rest. The protocol is designed around the assumption that roughly half the photons will be discarded. Alice sends twice as many photons as she ultimately needs for the key.

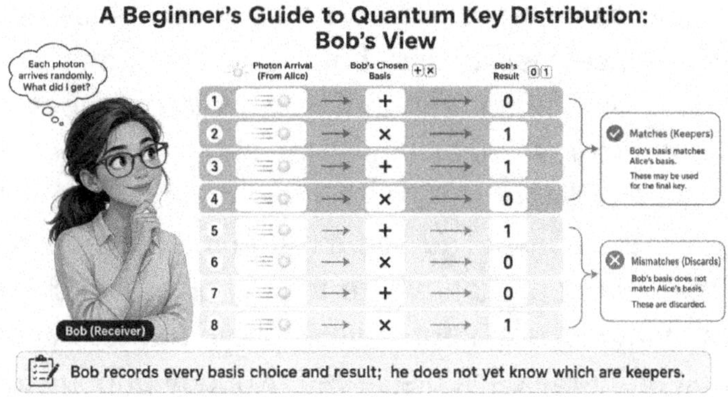

Diagram 4.6: Bob's full measurement log

There is one more physical detail worth noting here. Real detectors are not perfect. Fiber connections lose some photons along the way (the technical term is "loss"), and detectors sometimes fail to register a photon that actually arrived ("missed detections"). A real BB84 system accounts for these imperfections in the error-checking step. For now, assume that every photon Alice sends arrives and is detected. The idealized version captures all the important logic; the engineering adjustments come later.

Step Three: Sifting the Key

Once Alice has finished sending and Bob has finished measuring, the two of them establish a classical channel for the first time. This conversation is called sifting. Its purpose is to strip away all photons for which Alice and Bob used different bases, leaving only the shared subset where their measurements are guaranteed to agree.

The sifting conversation follows a strict rule: Alice and Bob announce their bases, but they never announce their bits. Alice might say that, for photon number 1, she used the rectilinear basis; for photon number 2, she used the diagonal basis; and so on down the list. Bob does the same. They go through the list together and mark every photon where their choices matched. Those photons are kept. Every photon where they used different bases is discarded. Nobody announces what their bit value was for any photon at any point in this step.

Why is it safe to announce the bases publicly? Because knowing which basis was used tells you nothing about the bit value unless you already measured the photon in the right basis. An eavesdropper who learns "photon 47 was sent in the diagonal basis" still has no idea whether it was a 0 or a 1 unless they measured it themselves. And if they measured it, they had to guess the basis before they knew it, and they had a 50 percent chance of guessing wrong. We will come back to why this creates a detectable fingerprint. For now, the key point is that announcing bases does not leak bits.

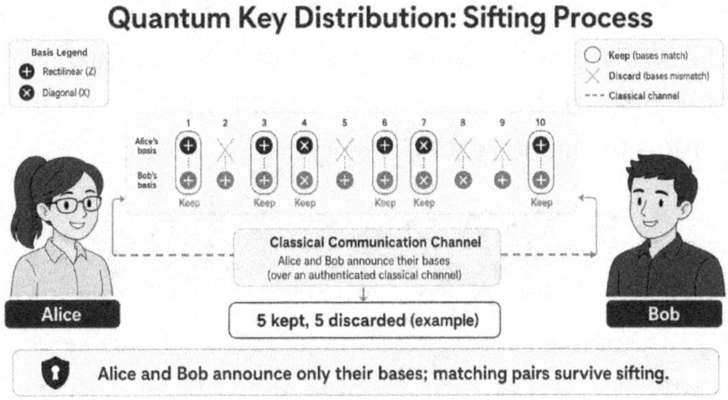

Diagram 4.7: The sifting conversation

After sifting, Alice and Bob each have a shorter list of bit values. Because they both used the same basis for every surviving photon, and because a photon measured in its own basis always yields the correct result, Alice's list and Bob's list should be identical. They should have arrived at the same string of bits without ever directly transmitting those bits. This is the proto-key, sometimes called the "sifted key."

The sifted key is longer than the final key will be, because the next step will trim it further. But already something remarkable has happened. Two people in different buildings have produced matching sequences of random bits that no one else was present to see. The bits were never sent. They arose from independent random choices colliding with the laws of quantum mechanics. That is the core miracle of QKD.

How long is the sifted key relative to Alice's original transmission? If Alice sends 10,000 photons, roughly 5,000 will survive sifting (the half in which the bases match). In practice, fiber loss and detector imperfections further reduce this, but the protocol is designed to account for those losses. Alice starts with a longer stream to ensure that enough photons survive to produce the required key length.

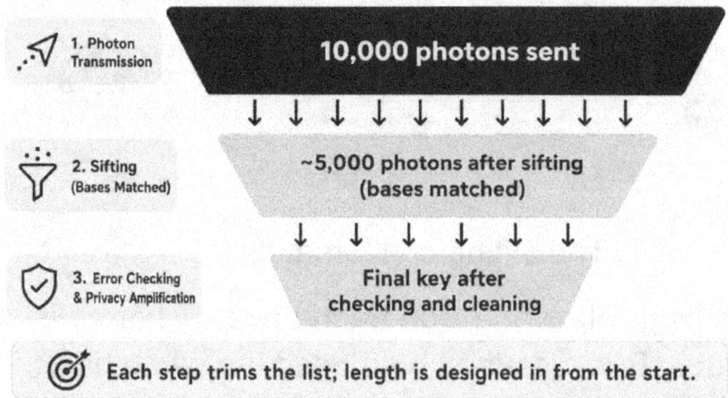

Diagram 4.8: From raw photon stream to sifted key

Step Four: Checking and Cleaning

Alice and Bob now have matching sifted keys, at least in theory. In practice, a few things can introduce disagreements. Detector imperfections produce occasional wrong readings. Fiber noise or temperature changes can flip a bit here and there. And, most importantly, an eavesdropper trying to intercept photons would introduce their own errors. The checking-and-cleaning step handles all of these sources of disagreement in two sub-steps: error estimation, followed by error correction plus privacy amplification (a way of shrinking the key to remove any partial information an eavesdropper might have gained).

The first sub-step is error estimation. Alice and Bob sacrifice a small portion of their sifted key for a public comparison. They announce (on the classical channel) the actual bit values for, say, 10 percent of their list. They pick these sacrifice bits at random. Then they compare the two lists for this sample. If Alice's bits and Bob's bits in the sample agree almost perfectly, the error rate is low, and they can be confident the channel was clean. If they find a large number of disagreements, it indicates that the photons have been disturbed, and they should be suspicious that an eavesdropper was involved. If the error rate is too high, they abort the exchange entirely and start over on a different day or a different channel.

Why does an eavesdropper cause errors? This brings us back to the measurement rule from Chapter 3. Remember: measuring a quantum state disturbs it. If an eavesdropper (let us call her Eve, the traditional third character) intercepts a photon to measure it, she faces the same problem Bob faces: she does not know which basis Alice used. She has to guess. Half the time she guesses wrong, and when she does, she sends Bob a photon that has been "re-polarized" in the wrong direction. When Bob then measures this photon in the correct basis, he gets a random result instead of the correct one. Eve's presence leaves a statistical fingerprint in the error rate. The more aggressively Eve intercepts, the higher the error rate climbs.

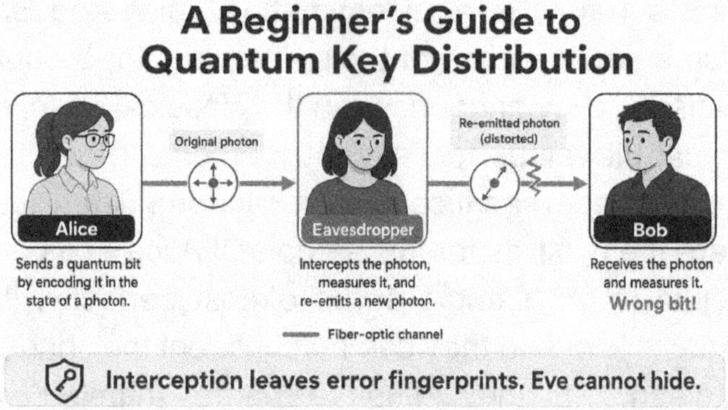

Diagram 4.9: How Eve's interception creates errors

The second sub-step is error correction. Alice and Bob use the remaining (non-sacrificed) bits, along with messages over the classical channel, to find and fix the mismatches in their lists. There are several clever algorithms for doing this efficiently. Still, the conceptual idea is straightforward: they compare groups of bits at a high level (announcing sums or parities of blocks rather than the bits themselves) and narrow down where the errors are until the two lists are identical. At the end of error correction, Alice and Bob have matching strings again.

But error correction leaks a little information. Every time Alice announces the parity of a block of bits, she hands Eve a tiny clue about the key. After error correction, Eve might know, in a statistical sense, a small fraction of the key bits. Privacy amplification is the final cleaning step that eliminates this residual risk. Alice and Bob apply a mathematical compression function to their corrected key, producing a shorter key that is provably free of any partial information Eve could have collected. Think of it as boiling down a large pot of vegetable soup into a concentrated cube: the result is smaller, but every drop of it is pure. The final key is shorter than the sifted key, shorter even than the error-corrected key, but it is clean.

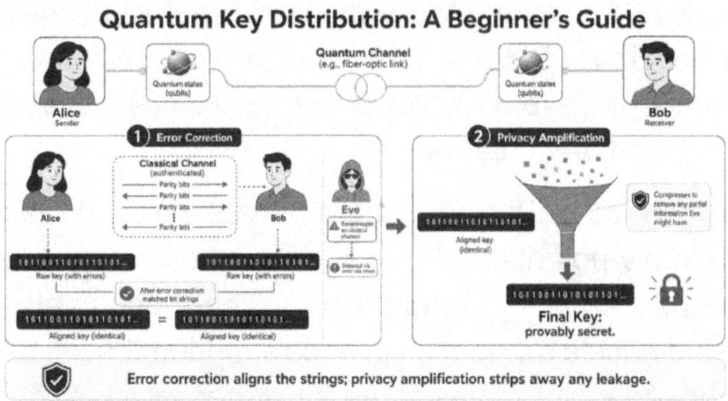

Diagram 4.10: Error correction and privacy amplification

The sacrifice bits used for error estimation are permanently discarded. The bits used in the parity announcements for error correction effectively give Eve a small amount of information, so they are compressed away in privacy amplification. The final key is therefore shorter than any of the intermediate strings. A rule of thumb for a clean, low-noise channel: Alice might send 100,000 photons to produce a final key of 10,000 bits. On a noisier channel, the ratio is less favorable. The protocol always adapts to deliver a key whose security is guaranteed regardless of the starting conditions, provided the error rate remains below a certain threshold (around 11 percent for a basic BB84 exchange; more sophisticated variants raise this threshold).

Why the Key Is Built, Not Sent

Now step back and look at the whole process with fresh eyes. Did Alice ever send a message that said: "here is the key"? No. Did Bob ever receive a packet labeled "key material"? No. Did any bit of the final key travel, in any form, on any channel? No.

What traveled on the quantum channel was a stream of randomly polarized photons. Each photon carried a randomly chosen orientation. The polarization encoded Alice's random bit in Alice's random basis, but without knowing the basis, the bit is inaccessible. The photons were raw material, not a finished product.

What traveled on the classical channel was basis announcements and error-estimation samples. Neither carries key bits. An eavesdropper who captured every message on the classical channel and every photon on the quantum channel (but could not measure photons without disturbing them) would end up with less information than Alice or Bob.

The key materialized only when Alice and Bob privately compared their results and identified the photons they both measured in the same basis. That identification happened in their own private logs, not on any channel. The key never "traveled." It crystallized from a shared physical experience: two people measuring the same quantum particles with matching tools.

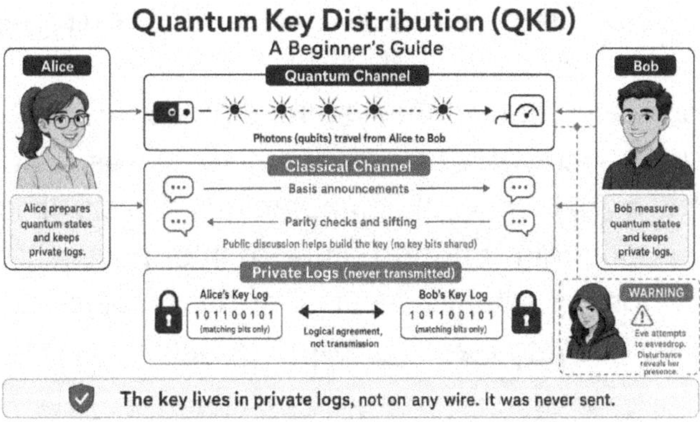

Diagram 4.11: Where the key actually lives

This is why the no-cloning rule matters so much. If Eve could make a perfect copy of each photon before it reached Bob, she could measure the copy at her leisure, wait to hear the basis announcements on the classical channel, and then reconstruct the key. But perfect copying of an unknown quantum state is forbidden by the laws of physics. Eve cannot copy the photon. She must measure it, and measuring it disturbs it. Her presence becomes visible in the error rate. Alice and Bob can detect her and walk away before she learns anything useful.

This is also why the measurement rule matters. A photon does not "carry" a definite bit in the way a letter carries words. The bit is defined only at the moment of measurement, in the context of the chosen basis. Before Bob measures in the correct basis, the bit does not fully exist as a classical value. There is nothing for Eve to steal because the stolen item does not solidify until the measurement happens. By the time Bob measures it, the moment of vulnerability is past.

The combination of these two rules (no cloning and measurement-induced disturbance) creates an eavesdropping detection system that no amount of technological cleverness can circumvent. This is not a lock that a fast enough computer might pick. It is a physical constraint on the nature of information itself. That is the qualitative shift that QKD represents compared to every classical key-distribution method.

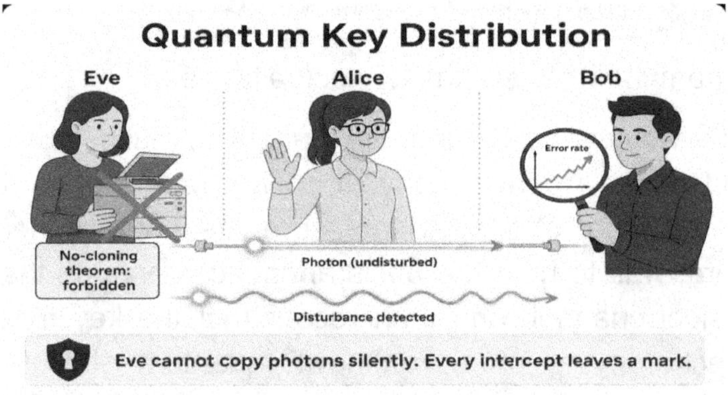

Diagram 4.12: The no-cloning shield

The Full Picture at Once

You have now seen each step in isolation. Let us run through the whole sequence one final time as a continuous story, so the pieces connect into a single coherent picture.

Alice sits at her transmitter and begins generating photons. For each photon, she flips two virtual coins: one for the bit (0 or 1) and one for the basis (rectilinear or diagonal). She orients the photon accordingly and fires it into the fiber. She logs her choices. She does this for every photon in the stream.

Bob sits at his detector and measures each photon as it arrives. For each photon, he flips his own virtual coin to pick a basis. He records the basis he chose and the bit value his detector returned. He does not contact Alice during this phase.

When the transmission is done, Alice and Bob open the classical channel. They go through their lists and announce, one by one, which basis they each used for each photon. When their bases match, they mark that photon as a keeper. When they do not match, they discard it. After this sifting conversation, both of them circle the same subset of photons in their logs. The bit values attached to those circled photons, the values they recorded during the measurement phase, are their shared sifted key.

Next, they sacrifice a sample of the sifted key to estimate errors. They announce those bit values publicly and compare. If the error rate is comfortably low, they proceed. If it is too high, they abort. Assuming they proceed, they run an error-correction protocol over the remaining bits, exchanging carefully designed parity information on the classical channel until their lists are provably identical.

Finally, they apply privacy amplification to the corrected list, compressing it into a shorter but guaranteed-clean final key. Each independently computes this compression and arrives at the same result. Nobody announced the final key. Nobody sent it. It exists only in Alice's and Bob's memories, arising from the random intersection of their choices and the physics of quantum measurement.

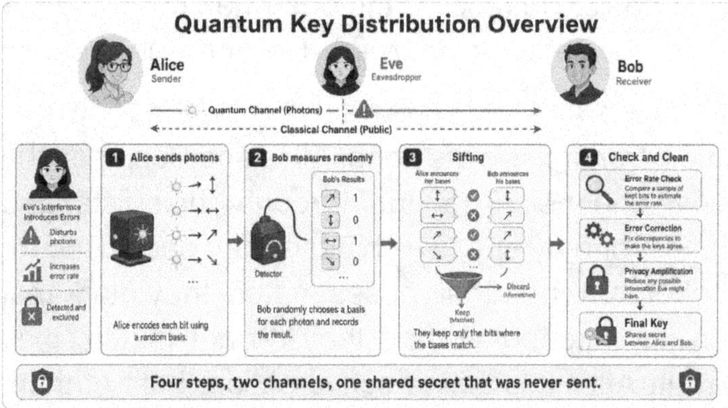

Diagram 4.13: BB84 end-to-end flow

This full picture reveals something that might seem counterintuitive at first: the more photons Alice sends, the longer and stronger the final key can be, but the relationship is not one-to-one. Because of sifting (which keeps roughly half), error estimation (which sacrifices a sample), and privacy amplification (which compresses the result), the final key is considerably shorter than the original transmission. This is not a flaw. It is the cost of security. Every bit trimmed along the way is a bit whose secrecy is guaranteed rather than assumed. The final key is small, clean, and provably private.

Modern QKD systems send millions of photons per second over fiber links. Even with all the trimming, they can produce keys at rates of tens of thousands of bits per second, fast enough for real-world encryption of voice calls, video streams, and data transfers. The physics does not change at any scale; only the engineering speeds up.

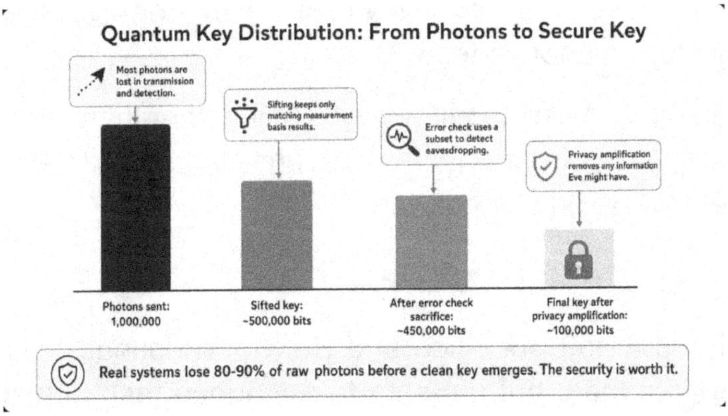

Diagram 4.14: Scale: from photon stream to working key

Key Concepts Checklist

• I can name Alice and Bob and describe their roles: Alice sends polarized photons; Bob measures them.

• I understand why QKD needs both a quantum channel (for photons) and a classical channel (for basis announcements and error correction).

• I can describe what Alice does for each photon: she randomly chooses a bit and a basis, orients the photon, sends it, and logs her choices privately.

• I understand why Bob's measurement is reliable when his basis matches Alice's and random when it does not.

• I can explain "sifting": Alice and Bob announce only their bases on the public channel, keep only matched-basis photons, and never announce the bit values.

• I understand the purpose of error estimation (detect eavesdroppers), error correction (fix residual mismatches), and privacy amplification (remove any partial information leakage).

• I can explain why the final key is built rather than sent: it materializes from private measurement records and never travels on any channel.

Questions to Explore

• Why is it safe for Alice and Bob to announce their bases out loud, but not safe to announce their actual bit values? What information does a basis reveal, and what does it withhold?

• About what fraction of Alice's original photons would you expect to survive all the way to the final key, and what happens to all the ones that are trimmed along the way?

• If the key is never sent over any channel, what exactly is traveling through the fiber? Is it meaningful to say the fiber "carries" information during the QKD exchange?

• How does the principle that "the key is built, not sent" defeat the eavesdropping threat that made Chapter 1's key-distribution problem so difficult?

Try This / Reflect

Grab a friend, two sheets of paper, and a coin. Agree that heads means bit 1 and tails means bit 0, and that one kind of flip means the rectilinear basis and another kind means the diagonal basis (you can use a regular coin for the bit and a different object for the basis, or use two separate coin tosses). One of you plays Alice. For each of ten "photons," Alice secretly flips twice: once for the bit and once for the basis. She writes both down but shows only the photon number. Bob also secretly flips once for his basis and writes it down. After all ten rounds, Alice and Bob compare only their basis choices, out loud. Circle the rounds where they matched. For those rounds, compare the bit values. You will discover that the matching rounds give Alice and Bob identical bit values that neither of them chose deliberately. That small collection of agreed-upon bits, perhaps four or five out of ten, is your hand-crafted BB84 sifted key. Notice how the key was never written on any note that was passed between you.

If you have more time, look at Appendix B, which walks through a complete numerical BB84 example with ten photons from start to finish, including the sifting step and a sample error check. Seeing the numbers fills in the concrete texture that this chapter keeps abstract by design.

Glossary Terms

• **Alice and Bob -** The traditional names for the two parties in a QKD exchange. Alice sends; Bob receives. They represent any two parties who want to share a secret key.

• **BB84 -** The original quantum key distribution protocol, named after its inventors Charles Bennett and Gilles

Brassard and the year it was published (1984). It is still the most widely taught QKD protocol.

• **Quantum channel -** The physical path, usually an optical fiber, that carries single photons from Alice to Bob. Quantum rules (no cloning, measurement disturbance) make it impossible to tap this channel silently.

• **Classical channel -** An ordinary communication link (internet, phone, or similar) that Alice and Bob use to compare bases and run error correction. It is assumed to be fully public; anyone may listen.

• **Basis -** A pair of orientations used to encode and measure photon polarization. BB84 uses two bases: rectilinear (vertical/horizontal) and diagonal (45-degree angles). Only measurements made in the same basis as the encoding yield the correct bit.

• **Sifting -** The step in BB84 where Alice and Bob publicly compare their basis choices and discard all photons where their choices did not match. After sifting, the remaining bit values are guaranteed to agree.

Key Takeaways

• BB84 uses two random choices per photon (bit and basis) to build a shared secret from nothing more than independent coin flips and quantum measurement.

• QKD requires both a quantum channel to carry photons and a classical channel to carry basis announcements; neither channel alone is sufficient.

• Alice and Bob reveal only their bases on the public channel, never their bit values; knowing the basis without the bit reveals nothing useful to an eavesdropper.

• Sifting discards roughly half the photons (those with mismatched bases), error estimation sacrifices a further sample to detect eavesdropping, and privacy amplification compresses the remainder into a clean final key.

• The final key is built from the intersection of two private measurement records and is never transmitted in any form, so there is nothing in transit for an eavesdropper to steal.

5 Catching the Eavesdropper

Overview

Chapter 4 showed how Alice and Bob build a shared secret key from photons. They send light, measure it, compare notes in public, and end up with a string of secret bits that neither of them ever said aloud. That is a remarkable achievement. But a skeptical reader is right to ask the next question: what if someone else was listening the whole time? In every form of classical communication, a skilled attacker can tap a wire or intercept a signal without the participants ever finding out. Silent eavesdropping is the default assumption in security design. It is precisely what quantum key distribution (QKD) breaks, and this chapter is dedicated to showing you exactly how and why.

The chapter introduces Eve, the traditional name for the eavesdropper in cryptography. Eve is not a passive observer who can hover nearby and absorb information. In QKD, to learn anything about the photons, she must measure them, and measuring them changes them. The no-cloning rule from Chapter 3 closes the most obvious escape route: Eve cannot copy the photons and measure the copies at leisure. When she guesses the wrong measurement basis, as she must do roughly half the time, she sends Bob a corrupted photon. Those corrupted photons show up later as unexpected disagreements when Alice and Bob compare a sample of their bits. The disagreement rate is a tripwire. If it climbs above a safe threshold,

Alice and Bob discard the key and start again, having lost nothing but time. The secret was never fully formed, so there is nothing to steal. This chapter will walk through every step of that logic in plain language and end with an honest conclusion: the ideal protocol is elegant and provably secure, but real hardware introduces complications that later chapters will address.

Opening Scenario

Imagine a city courier named Eve who has been offered a large sum to deliver a copy of the sealed documents Alice is sending to Bob. With ordinary paper documents, she would open the envelope, photograph each page, seal it back up, and deliver the original on time. Bob would never know. She has done it a hundred times. The trick works because paper does not care whether you look at it.

Today the documents are different. Each page is written in ink that exists in two possible orientations simultaneously, and opening the envelope forces every word to snap into one orientation or the other. Eve can read the result, but it is no longer the same document Alice sent. When she reseals the envelope and delivers it, some of the wording is subtly wrong. Alice and Bob agreed before the delivery to compare the first dozen words of their copies by phone. When those dozen words do not match the way they should, Bob stops reading. He does not know who tampered with the delivery, but he knows with near certainty that someone did. He calls Alice, and they agree to send a fresh document by a different route.

That is the heart of QKD security. Eve is not defeated because she lacks cleverness or computing power. She is defeated because the laws of physics prevent her from reading without changing, and because Alice and Bob have designed a tripwire that reveals the change. The rest of this chapter unpacks the physics and the procedure in enough detail that you can explain it to anyone.

What This Chapter Will Cover

• **Meet Eve, the Eavesdropper** - Introduce the third character, her goal, and why the standard eavesdropping tricks of classical networks fail when single photons are involved.

• **Why She Cannot Just Copy the Photons** - Apply the no-cloning rule from Chapter 3 to show that the most obvious attack, copying photons in transit and measuring the copies later, is forbidden by physics.

• **Why Measuring Betrays Her** - Show in step-by-step detail how Eve's forced measurement disturbs photons, how wrong-basis guesses corrupt the photon she forwards to Bob, and why she cannot avoid making wrong-basis guesses.

• **The Tell-Tale Error Rate** - Explain how Eve's interference raises the fraction of disagreements between Alice and Bob above what honest background noise would cause, and how that fraction becomes a security signal.

- **The Sampling Check** - Walk through the procedure Alice and Bob use to estimate the error rate by publicly comparing a randomly chosen sample of their key bits, then explain why a clean sample gives them justified confidence in the rest.
- **Abort and Retry, Lose Nothing** - Explain the graceful failure mode: if the error rate is too high, the key is discarded. Unlike a classical breach, a detected attack under QKD incurs only a time cost, not the loss of the secret itself.
- **Honest Caveat: Ideal vs. Real** - Distinguish the beautiful security guarantee of the idealized protocol from the messier real-world picture where hardware imperfections, side channels, and implementation flaws have been exploited. Set up the deeper treatment in later chapters.

Meet Eve, the Eavesdropper

Every story needs a third character. In the cryptography literature, that character has been called Eve for decades. The name is not accidental. It comes from "eavesdrop," the old image of someone standing under the eaves of a house to hear conversations through a window. Eve's goal is simple: she wants to learn the key that Alice and Bob are building so she can read the messages they will eventually encrypt with it. She wants to do this invisibly. A perfect eavesdropper leaves no sign of her presence.

In classical networks, Eve has a rich toolkit. She can tap a copper wire and copy the electrical signals without interrupting them. She can intercept wireless broadcasts without touching the transmitter or receiver. She can compromise a router and log every packet that flows through it. All of these attacks share one feature: the data being stolen is made of information that can exist in as many copies as you like. A copy of a digital file is indistinguishable from the original. Copying is free, silent, and leaves no physical trace on what was copied.

This is the default world that security engineers have designed against for the past half-century. Encryption exists precisely because we assume Eve can see every bit in transit. The strategy is not to hide the data from her eyes, but to lock it with mathematics so tightly that even if she has a perfect copy, she cannot read it without a key she does not have.

QKD changes the premise entirely. The channel carrying the key material is not a wire full of classical bits. It is a beam of individual photons, each one encoding one bit of potential key material in its polarization. And a photon is not a classical object. It does not have the comfortable property of being copyable. To learn its polarization, Eve cannot simply intercept it, measure it without disturbing it, and pass a perfect replica onward. Physics does not allow any of those three steps to happen silently.

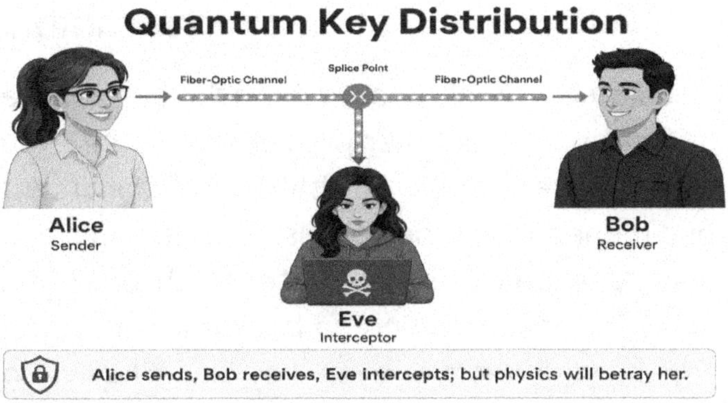

Diagram 5.1: Meet the three characters

Alice and Bob know Eve might be out there. They built that assumption into the design from the start. The protocol's security does not rest on Eve's absence; it rests on the fact that any action she takes will leave a detectable mark. The question is how that mark gets made and how Alice and Bob find it.

It helps to picture what Eve can and cannot do. She has access to the fiber between Alice and Bob. She can cut it, insert her own equipment, and reconnect it so perfectly that no physical inspection would reveal the splice. She can intercept every photon Alice sends. She has a quantum measurement device as good as Bob's, possibly better. She can send replacement photons toward Bob. She has unlimited classical computing power. She has all the time she needs. The only thing she cannot do is violate the laws of quantum physics. Those laws, it turns out, are enough to stop her.

Why She Cannot Just Copy the Photons

If Eve could copy each photon perfectly as it passed through her splice point, her problem would be trivial. She would make a copy, let the original continue to Bob, and measure her copy at leisure. She could even wait until Alice and Bob announce which basis they each used for each bit, then go back and measure in the same basis to recover every key bit perfectly. Bob would receive undisturbed originals. Alice and Bob would see no errors. Eve would have the entire key. Game over.

This is not idle speculation. It is precisely the attack that motivated one of the most important theorems in quantum information science, the no-cloning theorem, which Chapter 3 introduced. The theorem says that it is impossible to make a perfect copy of an unknown quantum state. Not difficult. Not currently impractical. Impossible, as a consequence of the same linear structure of quantum mechanics that underlies superposition and measurement.

Here is the intuition without any algebra. To copy something, you need to read it first, extract its exact state, and write that state into a blank. For a classical bit, reading is trivial because the bit is either definitely 0 or definitely 1. But a photon's polarization can be in a superposition of orientations until you measure it. If you measure it, you force it to pick a single orientation, and it loses its superposition. You now have a definite polarization, but it may not match the original superposition. You cannot copy what you have not yet measured, and measuring destroys the thing you wanted to copy.

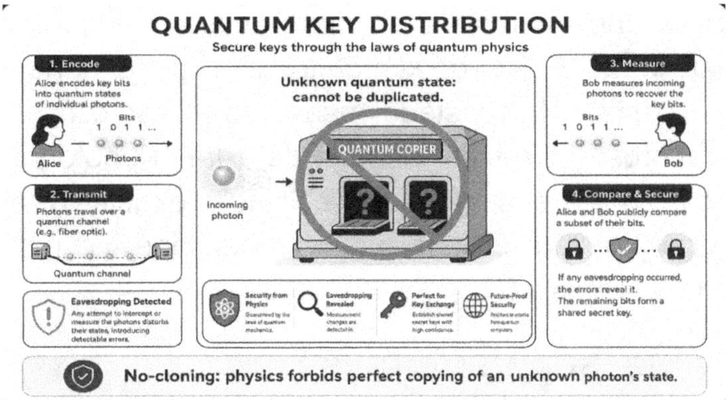

Diagram 5.2: The no-cloning wall

A more careful version of the argument goes like this. Suppose you had a device that could clone any quantum state. You feed it a photon in state A, and you get two photons both in state A. Now suppose state A is a superposition of two basis states, say, half horizontal and half vertical. The cloning device would need to produce two photons, each in that exact superposition. But quantum mechanics requires that any such device produce outputs that violate linearity; linearity is not optional. It is baked in at the deepest level. The no-cloning theorem is a mathematical proof, not an engineering limitation. No future technology will circumvent it.

What this means for Eve is that the most natural and powerful attack she could imagine is simply unavailable. She cannot be a passive observer who copies the signal and studies the copies. Every attempt to learn about the photons requires her to interact with them directly. And interaction, as we will see, leaves marks.

There is one loophole that might occur to a careful reader: what if Eve does not try to copy a perfect quantum state, but instead measures in some basis and records her result, then sends a fresh photon in whatever state she measured? That is exactly what she does try. But it does not save her, and the next section explains why.

Why Measuring Betrays Her

Let us follow Eve through her attack step by step. She has her splice in the fiber. Alice sends a photon polarized in one of four possible orientations, as in the BB84 protocol from Chapter 4. The four orientations are organized into two bases: the rectilinear basis (horizontal and vertical) and the diagonal basis (forty-five degrees and one-hundred-thirty-five degrees). Alice knows which basis and which orientation she chose. Eve does not.

Eve must choose a measurement basis before she measures. She cannot measure in both at once, and she cannot wait to see what Alice chose before deciding, because Alice does not announce that publicly until after the key exchange is complete. So Eve guesses. She picks a basis at random, measures the photon, records the result, and then sends Bob a fresh photon polarized according to whatever she measured.

When Eve guesses the correct basis, things look fine from her perspective. She measures the right property, gets the right result, and sends Bob a photon that matches what Alice sent. Bob measures it, gets the same result Alice intended, and no damage is done for that particular bit.

But when Eve guesses the wrong basis, something goes wrong at a fundamental level. Suppose Alice sent a photon polarized horizontally, which, in the rectilinear basis, reads as 0. Eve guesses the diagonal basis instead. In the diagonal basis, a horizontal photon is a superposition of diagonal and anti-diagonal, with equal probability of each. Eve's measurement forces the photon to "choose" one of those diagonal outcomes at random, say diagonal. Eve now holds a photon that has been forced into the diagonal orientation and records a diagonal result. She sends Bob a diagonally polarized photon.

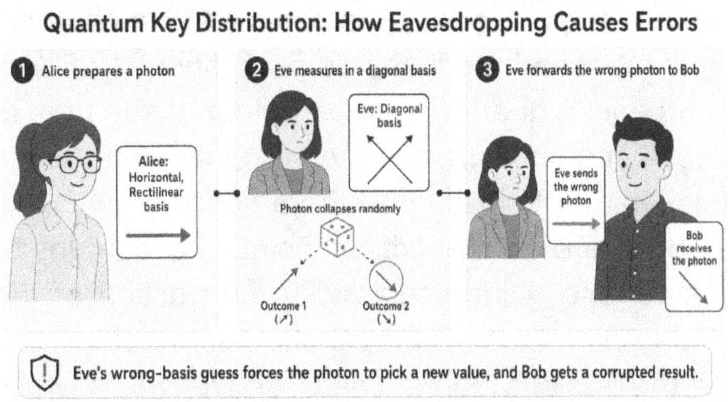

Diagram 5.3: Eve's wrong-basis guess

Now Bob measures. Suppose Bob, like Alice, happens to choose the rectilinear basis for this photon. A diagonally polarized photon measured in the rectilinear basis is, again, a superposition, and it will produce horizontal or vertical with equal probability. So Bob has a 50% chance of reading 1 when Alice sends a 0. When Alice and Bob later compare this bit during the sifting stage, they will see a disagreement where there should have been agreement.

The arithmetic compounds quickly. Eve makes a wrong-basis guess on half of all photons she intercepts, roughly speaking. For each wrong-basis photon, she has introduced a fifty percent chance of a detectable error at Bob's end. Working through the probability, Eve's intercept-and-resend attack on every photon raises the error rate between Alice and Bob to about twenty-five percent on the bits where their bases matched, compared to nearly zero in a clean line. That is a large and conspicuous signal.

Eve might try to intercept only a fraction of photons to reduce her footprint. If she intercepts half the photons, the error rate she introduces drops to about 12.5%. If she intercepts a quarter, it drops further. If she tries to intercept just a small fraction, the error she introduces becomes harder to distinguish from ordinary hardware noise. But here is the critical point: the amount of information she gains scales with the fraction of the signal she intercepts. The less she intercepts to stay hidden, the less she learns. There is no regime in which she can gain useful key information without introducing a detectable signal. The physics of measurement ensures it.

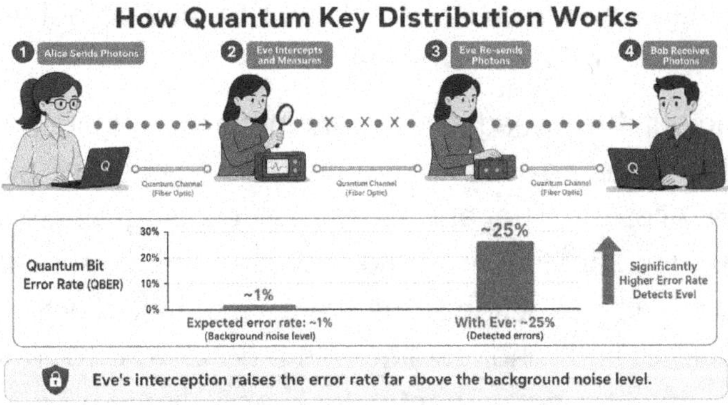

Diagram 5.4: How Eve's attack creates errors

It is worth pausing to appreciate the elegance of what just happened. The security argument does not rely on Eve being incompetent. It does not rely on her equipment being inferior to Alice's and Bob's. It does not depend on her not knowing the protocol. She can know everything about how BB84 works, have the best possible measurement devices, and operate with perfect efficiency. She still cannot avoid introducing errors, because errors arise not from her ignorance but from the fundamental way quantum measurement works. Measuring something changes it. Eve is not caught because she made a mistake. She is caught because the universe makes her make mistakes.

The Tell-Tale Error Rate

Alice and Bob emerge from the quantum channel transmission with two things: their own records of what they sent and measured, and the results of the public basis-matching conversation they had to sift down to the raw key. At this point they have a string of bits that, in a clean run, should be identical. But they do not yet know whether the run was clean.

To find out, they need to estimate the error rate on their raw key. The error rate is simply the fraction of bits where Alice's value and Bob's value disagree. In a perfectly clean line with no eavesdropping and no hardware imperfections, that rate would be zero. In practice, even an honest line has a nonzero error rate due to detector noise, photon loss, and alignment drift. This honest background noise is typically well below 1%. Eve's intercept-and-resend attack, as we saw, pushes the rate toward twenty-five percent. Even a partial attack pushes the rate above the background.

The error rate is therefore a security signal. Not a perfect signal: there is no way to distinguish every photon that Eve corrupted from every photon that honest hardware degraded. But the total rate tells a story. Alice and Bob agree in advance on a threshold. If the error rate on the sample they test is below the threshold, they proceed. If it is at or above the threshold, they abort. Choosing the right threshold is a technical decision that the protocol's designers have studied carefully, and the details depend on the specific hardware and expected noise levels. For the purpose of understanding the concept, you can think of the threshold as a line drawn at the boundary between what honest noise can explain and what suggests interference.

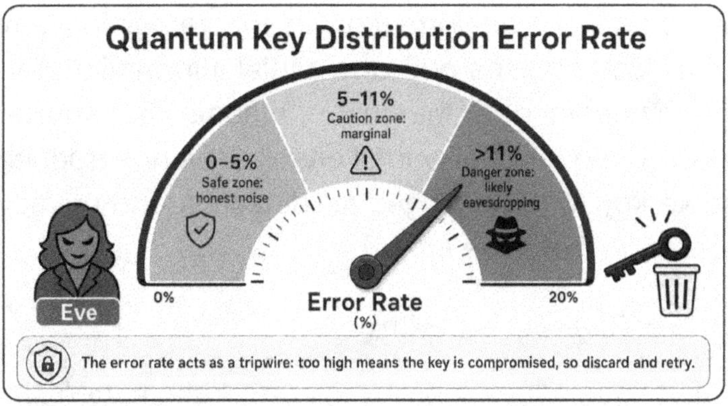

Diagram 5.5: The error rate as a tripwire

What happens to the bits that Alice and Bob use to estimate the error rate? Those bits are sacrificed. Alice and Bob announce their actual values for those bits over the public channel, compare them, count the disagreements, and then discard those bits. Because the values were announced publicly, they are no longer secret and cannot be part of the final key. This sacrifice is a deliberate trade: a relatively small number of bits is spent to buy confidence in the much larger remaining set. The protocol is designed so that the sample size is large enough to give statistically meaningful confidence without spending too many bits.

There is a subtler issue worth mentioning. What if Eve is clever enough to attack only some photons, just enough to gain a small amount of information while keeping the error rate below the threshold? This situation, sometimes called the partial information problem, is addressed by a step called privacy amplification, which Chapter 6 covers in more detail. For now, the short version is that Alice and Bob apply a mathematical transformation to the remaining key bits that "compresses out" any partial information Eve might have gained, at the cost of a somewhat shorter final key. Even a clever partial attack does not produce a usable key for Eve. Together, detection and privacy amplification close the loop.

The Sampling Check

Now that you understand why the error rate matters, let us walk through the actual procedure Alice and Bob use to measure it. This is called the sampling check, or sometimes the privacy test, and it is one of the most important moments in the whole QKD protocol.

After sifting their raw key bits from the photon exchange, Alice picks a random subset of those bits. She does this by generating a list of random positions, say positions 7, 12, 34, 91, and several hundred more, and announcing those positions over the public classical channel. Both Alice and Bob then read out their bit values at those positions over the public channel and compare them. They count the number of positions where their values disagree and divide by the total number of sampled positions. The result is their estimate of the error rate.

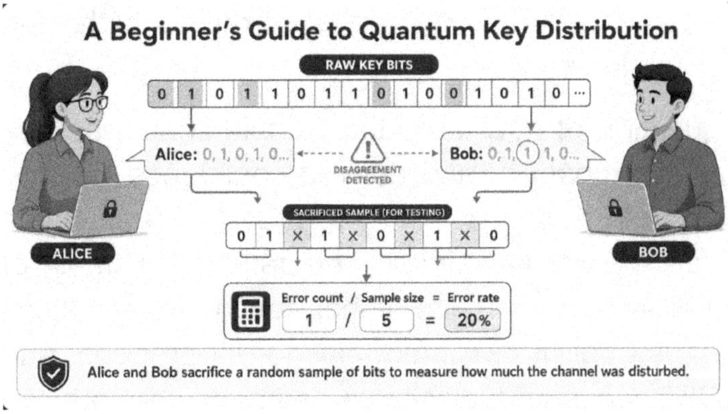

Diagram 5.6: The sampling check procedure

The randomness of the sample selection is important. If Eve somehow knew in advance which positions would be checked, she could refrain from attacking those specific photons and attack all the others freely. By choosing positions after the photon exchange is complete and by announcing them publicly only at check time, Alice and Bob ensure that Eve had no way to know which bits to leave clean. Any attempt by Eve to intercept photons is spread across the whole stream, and the sample is a fair representation of that stream.

The statistical logic here is the same logic used in opinion polling or product quality testing. You do not need to test every single item to know the overall quality; a sufficiently large random sample gives you a reliable estimate of the whole. If Alice and Bob sample a few hundred bits and find that only one or two are in disagreement, they can say with high confidence that the error rate across all the remaining bits is also very low. If they find that fifty of those bits are in disagreement, they know the error rate is high, and something is wrong.

The question of how large the sample should be is a matter of probability and desired confidence level. A sample of 100 bits might provide reasonable confidence for a low-stakes application. A sample of 1,000 bits provides much higher confidence. Protocol designers publish exact formulas for the tradeoff, and real QKD systems typically use samples that provide confidence levels measured in many standard deviations. The reader does not need to work through those formulas; the key insight is that the sample check transforms a physical signal, the disturbance Eve created, into a number that Alice and Bob can act on.

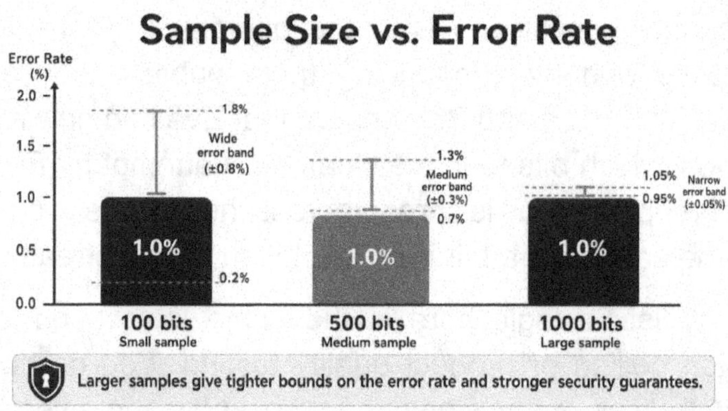

Diagram 5.7: How sample size affects confidence

One thing the sampling check does not do is tell Alice and Bob exactly where Eve's interference happened or how many photons she touched. It only tells them the aggregate error rate. That is enough information to make the pass-or-fail decision, and that is all the procedure needs to do. Alice and Bob do not need to track Eve down, prove her identity, or understand her equipment. They only need to know whether the channel was clean enough to trust.

After the check, the sampled bits are discarded regardless of whether any errors were found. The remaining unchecked bits form the candidate key, which is subject to the privacy amplification step mentioned above. If the error rate passed the threshold, that candidate key, after privacy amplification, is the final shared secret. It was never transmitted in its final form, and Eve, even if she intercepted many photons, ends up with a set of partial information that, after privacy amplification, is not statistically useful for breaking the key.

Abort and Retry, Lose Nothing

Let us consider what happens in the scenario every security professional dreads: the attack succeeds in the sense that it is detected. Alice and Bob finish their sampling check and find an error rate of, say, eighteen percent. Eighteen percent is well above any reasonable threshold. Something has disturbed the channel badly. They cannot know for certain whether the disturbance was caused by Eve, a hardware malfunction, or severe interference on the fiber, but they know the channel is not trustworthy for key generation today.

Under BB84 and its descendants, the response is simple: discard the key and try again. Alice and Bob throw away every bit from this session. They pick a new starting time, generate new random bases and bits, and run another session over the same or a different channel. If the problem was a transient hardware fault, the next run will probably be clean. If Eve is still present and attacking, the next run will also show high error rates. They can try as many times as they like. Each failed session costs them time but nothing else.

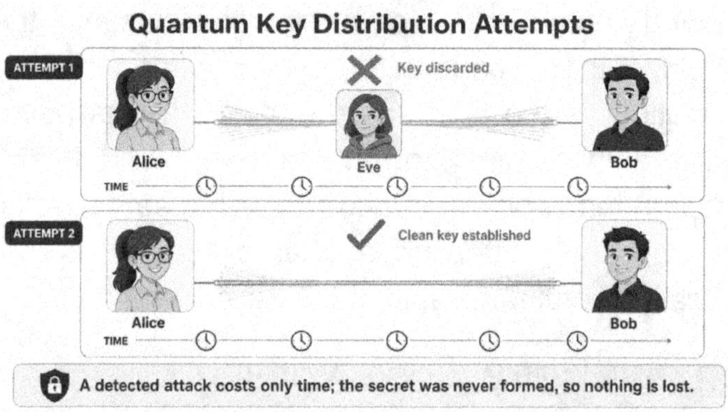

Diagram 5.8: Abort and retry

This graceful failure mode is one of the most striking features of QKD compared to classical security. Think about what happens when a classical system is breached. An attacker who intercepts an encrypted communication may not be able to read it immediately, but they have a perfect copy of the ciphertext, which they can study for years. If the key is later compromised, the old messages are retroactively vulnerable. This is the "harvest now, decrypt later" threat from Chapter 2. The harm happens silently, and the consequences may not be visible for years or decades.

In QKD, the key being built is not the same as the messages it will eventually protect. The key is a fresh, random string that will be used once and discarded. If the key generation session is compromised, the key is discarded before it is ever used to encrypt anything. There are no old messages to decrypt retroactively. There is no ciphertext for Eve to carry away. There is no accumulating vulnerability. Eve's effort, however sophisticated, produces nothing. She does not even gain the satisfaction of having a locked box to work on. The box was never built.

Furthermore, Alice and Bob's decision to abort is based on physics, not on a password being guessed or an alarm system being tripped. The tripwire is not something Eve can disarm with enough cleverness because it is built into the behavior of photons. She cannot make a measurement that is gentler than quantum mechanics allows. She cannot learn polarization without collapsing superpositions. She cannot avoid introducing errors without giving up on gaining information. The laws of physics have no bypass.

There is one important nuance. The abort-and-retry strategy assumes that Alice and Bob have not yet used the suspect key to encrypt sensitive messages. In a properly designed QKD system, the key is not released for use until the sampling check has passed. This sequencing is not a detail; it is central to the security model. Any implementation that uses key bits before the check is complete has broken the security argument. Well-designed QKD systems enforce this sequencing rigorously.

Diagram 5.9: Why classical breaches are worse

The retry logic also illuminates a practical aspect of real QKD deployments. Because sessions can fail for honest reasons, such as fiber degradation, detector noise spikes, or alignment loss, QKD systems need to handle retries gracefully without raising a security alarm every time. Operators learn the baseline error rate of their hardware and set thresholds that distinguish honest variation from true eavesdropping signals. Managing this calibration carefully is part of the engineering discipline that accompanies the theoretical security guarantees.

Honest Caveat: Ideal vs. Real

Everything in this chapter so far has described the ideal BB84 protocol operating under ideal conditions. Photon sources produce exactly one photon per pulse. Detectors have perfect efficiency and no dark counts. The fiber is perfectly characterized. Eve is limited to the intercept-and-resend attack. Alice and Bob perform all their steps with mathematical precision. Under these idealized assumptions, the security argument is complete and mathematically proven. It is one of the crown jewels of modern cryptography.

Real systems do not operate under ideal conditions. This is not a reason to dismiss QKD, but it is a reason to be clear-eyed. Several gaps between the ideal and the real have been identified and in some cases exploited by researchers.

The first and most famous gap involves photon sources. An ideal BB84 source produces exactly one photon per pulse. Real laser sources, when attenuated to very low intensity, sometimes produce pulses with two or three photons rather than exactly one. When a pulse contains two photons instead of one, Eve can grab one photon and let the other continue to Bob. She can measure the photon she grabbed after Alice and Bob publicly announce their bases, thereby learning the bit value without disturbing the photon Bob received. This attack, called the photon-number-splitting attack, was a serious concern for early weak-laser implementations of QKD. It does not violate fundamental physics, but it does break the assumption that every pulse is a single photon. The countermeasure, a protocol modification called decoy states, addresses this by having Alice randomly insert pulses of varying intensities, allowing her and Bob to test whether Eve is statistically picking off extra photons.

Diagram 5.10: The multi-photon loophole

The second major gap involves detectors. Strong light pulses can influence real single-photon detectors in ways that go beyond their normal operating mode. Researchers have demonstrated "blinding" attacks in which Eve floods a detector with bright light, forcing it into a regime in which it responds only to Eve's carefully timed trigger pulses. Under such an attack, the detector reports results that Eve controls rather than those corresponding to actual quantum states. The blinded detector still passes its internal self-checks, so Alice and Bob's system sees nothing unusual. This kind of side-channel attack does not break the mathematics of BB84; it bypasses the mathematics entirely by breaking an assumption the math relies on, namely that the detector faithfully reports quantum measurement results. Countermeasures include detection-device-independent QKD protocols and hardware-level defenses, both of which are discussed in Chapter 6.

A third gap involves information that leaks through channels other than the quantum channel itself. The timing of photon arrivals, the electrical signals in the control electronics, and even the sound of switching equipment can, in principle, carry information that a sufficiently sophisticated adversary could exploit. These are called side-channel attacks, and they are the bane of all physical cryptographic implementations, quantum or classical. They do not reflect a flaw in quantum mechanics. They reflect the difficulty of building physical hardware that perfectly embodies a mathematical abstraction.

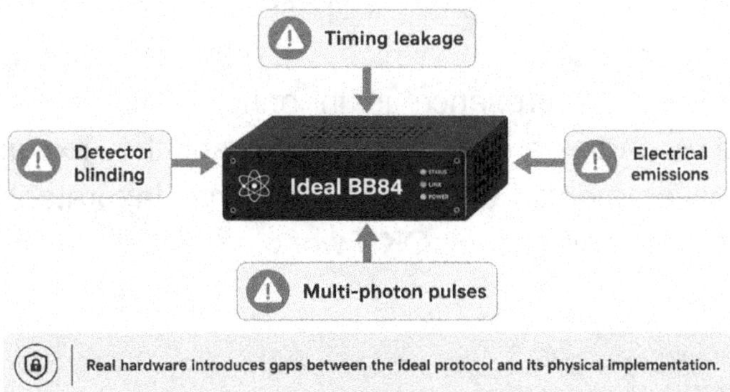

Diagram 5.11: Side channels and the ideal-real gap

What should you take from all this? The honest answer is that QKD's security guarantee is real and mathematically rigorous, but it applies to the idealized protocol running on idealized hardware. Closing the gap between that ideal and real deployments requires careful engineering, protocol extensions, and ongoing security research. The best commercial QKD systems today are much closer to the ideal than early prototypes were, and active research continues to close the remaining gaps. But anyone who tells you that a deployed QKD system is automatically and unconditionally unbreakable is overselling. The physics is ironclad; the engineering must be too. Chapter 6 will give you the tools to assess the engineering side. Chapter 10 will return to limitations with a broader critical lens.

The important point, which should not be lost in the honest caveats, is this: even with all the imperfections acknowledged, QKD is the only key distribution technology that derives its security from physical laws rather than computational hardness assumptions. It is the only approach for which we can say that an eavesdropper's presence is in principle detectable. Those properties are genuinely novel, and they explain why governments, banks, and national laboratories have invested in making QKD work in practice.

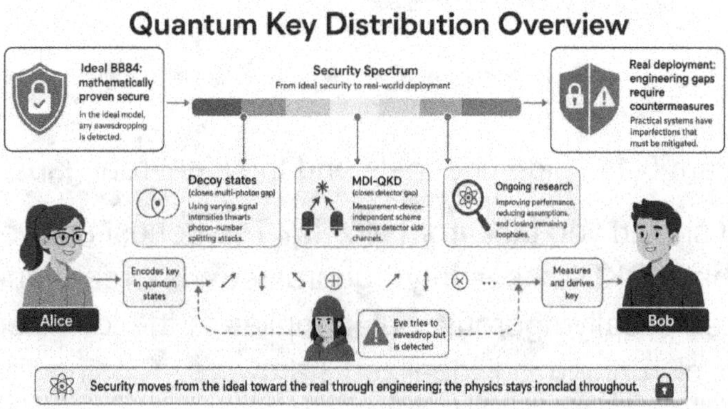

Diagram 5.12: The spectrum from ideal to real QKD

For the manager or decision-maker reading this book, the practical takeaway is that evaluating a QKD product means asking not just whether the protocol is sound, but also whether the implementation has addressed known hardware vulnerabilities. A QKD system built on well-tested protocols with decoy states, robust detector protections, and regular security audits is a meaningfully different thing from a system that implements only the bare theoretical protocol without engineering hardening. The difference matters.

Key Concepts Checklist

• I can explain who Eve is, what she wants, and why classical eavesdropping techniques fail against a quantum channel.

• I understand why Eve cannot copy a photon to measure at leisure, and which physical law prevents it.

• I can describe step by step how Eve's wrong-basis measurement corrupts the photon she forwards to Bob.

• I understand that Eve's interception raises the error rate between Alice and Bob by a predictable and detectable amount.

• I know how Alice and Bob use a randomly sampled subset of their key bits to estimate the error rate and detect eavesdropping.

• I understand why a detected attack under QKD costs only time and never exposes the secret itself.

• I can name at least two ways that real hardware deviates from the ideal protocol and one countermeasure for each.

Questions to Explore

• Why is "eavesdropping leaves a trace" such a radical departure from the assumptions that classical security systems are built on?

• If Eve gets lucky and guesses the correct basis for some photons, why does she still get detected overall when she attacks a significant fraction of the stream?

• What factors should an organization weigh when deciding where to set the error-rate threshold that triggers an abort? Who should make that decision, and what expertise do they need?

• How is the "detect and discard" response to an attack different from, and arguably superior to, a classical security model that tries to prevent eavesdropping at all costs?

Try This / Reflect

Extend the pencil-and-paper BB84 game from Chapter 4 by recruiting a third person as Eve. In this version, Eve sits between Alice and Bob. For each photon Alice sends, Eve must secretly write down a randomly chosen basis, a measurement result consistent with that basis, and then pass along a replacement photon in her measured state. Run the game twice without telling Bob whether Eve is present. After both runs, do the sampling check together by comparing a random handful of matched-basis bits out loud. Notice how the run with Eve produces noticeably more disagreements. Watching the error rate jump from near zero to something conspicuous is the most direct way to feel in your bones why QKD works. If you have a willing group, run it a third time with Eve intercepting only half the photons and see how the error rate drops but does not disappear.

You might also try this as a thought experiment. Imagine you are Eve. You have intercepted a photon that you suspect is polarized either horizontally or diagonally, but you do not know which. Ask yourself: is there any measurement you could make that would tell you the answer without changing the photon's state? Then ask whether there is any measurement you could make that would let you send Bob a perfect replica regardless of the answer. If you work through why the answer to both questions is no, you have understood the deepest reason why QKD security holds.

Glossary Terms

- **Eve** - The traditional name for the eavesdropper in cryptography thought experiments; in QKD, the party attempting to intercept the quantum channel without being detected.

- **Intercept-and-resend attack** - The attack strategy in which Eve measures each photon and sends Bob a fresh photon in the state she measured, necessarily introducing errors due to wrong-basis guesses.

- **No-cloning theorem** - The physical law stating that an unknown quantum state cannot be perfectly copied, which eliminates the most obvious eavesdropping strategy.

• **Error rate** - The fraction of key bits on which Alice and Bob's values disagree after basis sifting; the primary signal used to detect eavesdropping.

• **Sampling check** - The procedure in which Alice and Bob publicly compare a randomly selected subset of their key bits to estimate the error rate and decide whether to proceed or abort.

• **Threshold** - The agreed-upon maximum error rate below which Alice and Bob accept the key as secure; above the threshold, they discard the key and retry.

• **Privacy amplification** - A mathematical post-processing step that compresses the remaining key bits to remove any partial information Eve may have gathered through a limited interception, resulting in a shorter but more fully secret key.

Key Takeaways

• In QKD, eavesdropping cannot be silent: measuring a photon disturbs it, and that disturbance shows up as errors that Alice and Bob can detect.

• The no-cloning theorem blocks the most powerful possible attack by preventing Eve from copying photons and measuring the copies at leisure after learning the bases.

• Eve's intercept-and-resend attack raises the error rate between Alice and Bob by a predictable and substantial amount, providing a reliable security signal.

• Alice and Bob detect eavesdropping by publicly comparing a randomly sampled subset of their key bits;

a clean sample gives statistically strong confidence in the rest.

• A detected attack under QKD costs only time, not secrecy: the key is discarded before use, leaving Eve with nothing, unlike classical breaches that expose data retroactively.

6 From Idea to Hardware

Overview

Quantum key distribution (QKD) protocols look elegant on paper. Two people exchange quantum states, compare a fraction of them, and end up sharing a secret key. The physics is real, and the logic is sound. But a protocol does not send your bank transfer. A machine does. This chapter moves from the world of ideas to the world of things: lasers, glass fiber, cooled detectors, and alignment circuits that work around the clock. The goal is not to turn you into an optical engineer. The goal is to replace the vague image of "a quantum device" with a clear, honest picture of what the hardware actually does and why each piece of it matters.

Along the way, the chapter keeps a promise the book made at the start: honesty about imperfections. Real photon sources sometimes fire two particles instead of one, and that gap creates a real attack. Real fiber swallows most of the light before it reaches the far end. A carefully crafted bright pulse can fool real detectors. Each imperfection has a counter-move, from the decoy-state trick to careful calibration to the emerging framework of device-independent QKD. The central lesson is this: QKD's famous security guarantees are proofs about ideal devices. Closing the gap between the ideal and the real is an engineering challenge, and engineering is where today's most serious security work is happening.

Opening Scenario

Marcus is a senior network engineer at a regional hospital group. His team has just taken delivery of a new QKD appliance, a rack unit about the size of a desktop computer, connected to a dedicated fiber between two of the hospital's buildings. The vendor's technician has finished the installation and is about to leave. Marcus stares at the front panel. There are three status lights, a small display labeled "QBER" (he will learn what that means), and a single fiber port. He asks the technician: "What's actually inside this thing?" The technician smiles, opens a side panel, and points to a tiny laser module, a series of small mirrors and crystals on a vibration-dampened platform, and a detector module with a cooling fan. "That's it," she says. "That and a lot of software keeping it all stable." Marcus thought quantum hardware would look more dramatic. It does not. It looks like very careful engineering.

That gap between expectation and reality is the theme of this chapter. QKD feels like physics magic, but it runs on hardware that obeys the same rules as any other precision instrument: it drifts, it loses signal, and it can be attacked if you know where to push. Understanding the hardware does not diminish the physics. It shows you where the physics lives and where the engineering takes over. By the end of this chapter, you will know the job of every major component, understand why each one poses a challenge, and see how clever people have solved, or are still solving, those challenges.

What This Chapter Will Cover

• **Making Single Photons** - how real systems try to emit one photon at a time, why that is harder than it sounds, and why a source that occasionally fires two photons at once creates a security hole.

• **The Decoy-State Trick** - how sending occasional pulses of different brightness lets Alice and Bob detect a specific attack, and how this simple idea restored confidence in real-world QKD.

• **Sending Photons Down Fiber** - how photons travel through glass, why the fiber absorbs most of them over long distances, and why loss is the practical enemy more often than eavesdropping is.

• **Detecting Single Photons** - the remarkable detectors that register the arrival of a single particle of light, why many must be cooled, and how a faint click becomes a recorded bit.

• **Keeping the System Aligned** - why real links must constantly recalibrate to fight temperature changes, vibration, and slow drift, and why a QKD appliance is as much a control system as an optical one.

• **When Hardware Can Be Fooled** - the documented cases where attackers defeated real QKD systems not by breaking the physics but by exploiting quirks in the hardware, and what that reveals about the nature of real-world security.

• **Why Engineering, Not Just Physics, Determines Security** - the chapter's main conclusion: the physics sets the ceiling, the engineering determines how close you get to it.

Making Single Photons

Picture a water faucet. If you want to deliver exactly one drop at a time, you cannot just leave the tap open. You need a mechanism to meter the flow. A photon source in a QKD system faces exactly that challenge. The protocol requires one photon per pulse. Two photons per pulse is a problem. Zero photons per pulse is a wasted clock cycle.

The most common approach in early- and mid-generation QKD systems is to use a highly attenuated laser. The word "attenuated" means "turned down very low." A normal laser fires a bright beam containing billions of photons per second. Engineers reduce that intensity with optical filters until, on average, a pulse contains well under one photon. The statistical result is that most pulses carry zero photons, a significant fraction carry exactly one, and a very small fraction carry two or more. This sounds acceptable but, as you will see shortly, that small multi-photon fraction is exactly what an eavesdropper can exploit.

The attenuated laser approach is practical and cheap. It is why most deployed QKD systems use it. But researchers have long pursued better alternatives. Quantum dots are tiny semiconductor structures that can be engineered to release exactly one photon when excited by an electrical pulse, much like a single-atom light switch. Color centers in diamond crystals, such as the nitrogen-vacancy center, behave similarly: hit the right atomic defect with the right energy, and it emits one photon. These are called "true single-photon sources," and they offer a cleaner solution. Their challenge is that they are currently more expensive, more temperature-sensitive, and harder to integrate into compact hardware. The field is advancing quickly, and future systems will likely rely on them more heavily.

There is also a middle path called spontaneous parametric down-conversion, or SPDC. A special crystal splits one high-energy photon into two lower-energy photons simultaneously. If you detect one photon, you know its partner exists. This pairing property is useful for certain advanced protocols. For now, the important point is that engineers have more than one option, and the choice of source shapes the system's security profile.

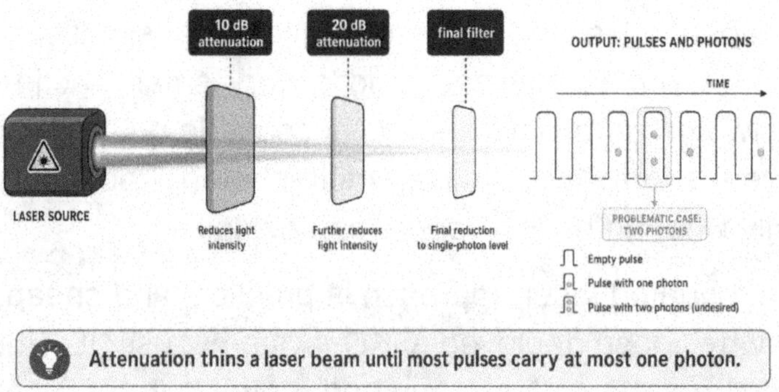

Diagram 6.1: How an attenuated laser becomes a photon source

Why does the two-photon case create a security risk? Think of it this way. If Alice sends a pulse containing two identical photons, an eavesdropper named Eve can intercept one of them, keep it, and let the other continue to Bob. Bob receives a photon and records a bit. Alice never knows one went missing because the pulse still "arrived." Eve now has a copy of the bit, encoded in the photon she captured. This attack is called the photon-number-splitting (PNS) attack, and it is one of the most important practical threats to early QKD systems. It does not break the quantum no-cloning principle because Eve is not copying a photon. She is simply stealing one of two identical photons that were already there. The distinction matters enormously.

The PNS attack shows that hardware imperfections are not just efficiency problems. They are potential security holes. The entire framework of QKD security assumes the source emits at most one photon per pulse. When that assumption fails, even partially, the security proof does not apply. Engineers needed a way to use the practical, affordable attenuated-laser source while still protecting against PNS attacks. The answer was the decoy-state method.

The Decoy-State Trick

Imagine you are a museum security guard trying to determine whether a visitor is sneaking objects out under their coat. One clever approach: sometimes send test visitors carrying nothing, sometimes send them carrying decoys of varying sizes, and watch whether the pattern of what exits the museum matches what you expected. If an actual thief is intercepting real objects and replacing them with look-alikes, the statistics on what gets through will not match those of a world with no theft. The thief's interference shows up as an anomaly.

The decoy-state method works on exactly this logic. Alice does not send all pulses at the same average brightness. Instead, she occasionally sends pulses with a slightly higher average photon number, and sometimes sends nearly empty "vacuum" pulses. Bob does not know, in advance, which pulses are which. Eve does not know either. After transmission, Alice and Bob publicly compare the statistics of how many pulses arrived, broken down by intensity level. If a PNS attack is underway, Eve selectively blocks low-photon pulses (where she cannot steal one and let another through) and allows high-photon pulses (where she can). That selective behavior leaves a statistical fingerprint in the arrival counts.

The key insight is subtle but important. In a world with no eavesdropping, the fraction of pulses that arrive successfully should follow a predictable pattern across intensity levels. It is governed purely by the fiber's loss characteristics. If Eve is interfering, she has to make choices: she can pass some pulses, block others, and store information she stole. Those choices distort the pattern. Alice and Bob compare the actual arrival statistics against the expected pattern and can compute a tight upper bound on how much information Eve could have gained. If that bound is small enough, they proceed with key distillation. If not, they abort.

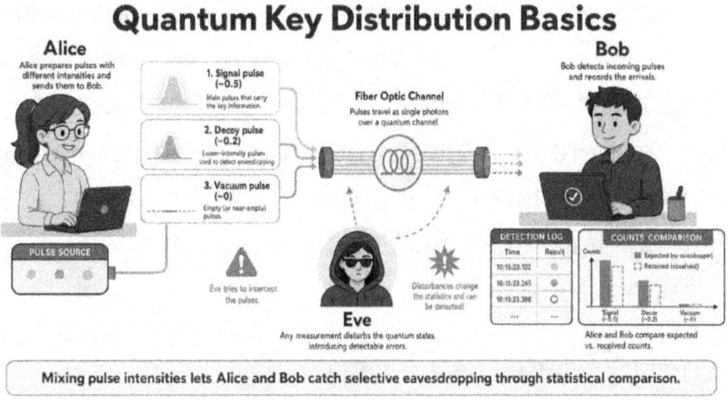

Diagram 6.2: Alice sending signal pulses and decoy pulses

The decoy-state idea was proposed in the mid-2000s, and it transformed the practical security landscape for QKD. Before decoy states, the PNS attack was a legitimate showstopper for attenuated-laser systems. After decoy states, the same attenuated-laser hardware became provably secure against PNS, provided the decoy analysis is done correctly. It is a beautiful example of a security technique built not on new physics but on clever statistics layered atop existing hardware. It also illustrates a recurring theme: real-world QKD security is not just about the quantum channel. It is about what you can prove from the data you collect.

Implementing decoy states requires that Alice's source can produce pulses of different intensities on command, switching rapidly and unpredictably. That sounds simple, but it means the laser modulation electronics must be precise, fast, and free of leakage. Any predictable pattern in which pulses are decoys and which are signals would hand Eve information she should not have. The randomness of the decoy schedule must itself come from a trustworthy random number generator. Once again, one engineering requirement connects to another. A QKD system is not a collection of independent parts. It is a chain in which every link matters.

Sending Photons Down Fiber

A standard optical fiber is a strand of extremely pure glass, roughly the width of a human hair, wrapped in protective cladding. Ordinary communications fibers carry signals at power levels of milliwatts or more. QKD fibers carry individual photons. The difference in scale is almost incomprehensible. A milliwatt is roughly a trillion photons per second. A QKD pulse might contain on average half a photon. The fiber does not care; it treats both the same way. But the consequences of that indifference are severe when your signal is already barely there.

The main problem is absorption. As a photon travels through the glass, there is a chance it will be absorbed by an impurity or scattered by a tiny structural irregularity. That chance compounds with distance. In modern telecommunications fiber at the wavelengths preferred by QKD systems, the signal drops by roughly half every 15 to 20 kilometers. Engineers express this in units of decibels per kilometer. A rule of thumb: every ten kilometers or so, roughly half the photons are gone. At fifty kilometers, only a few percent remain. At one hundred kilometers, the fraction is tiny. At two hundred kilometers, reaching Bob with any photons at all requires either extremely good equipment, a relay, or a lot of patience.

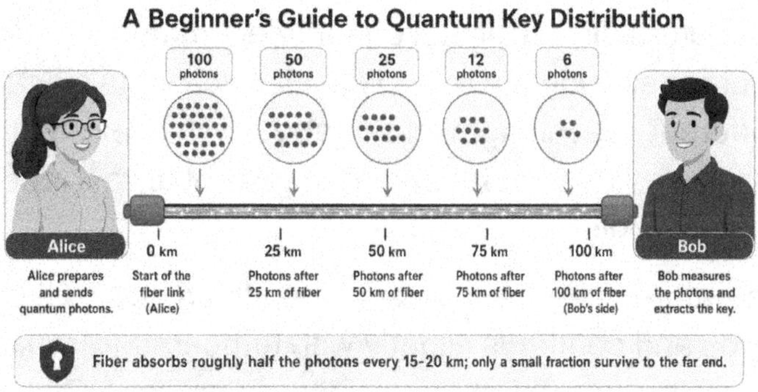

Diagram 6.3: Photon loss along fiber distance

This loss is not a security breach. It is physics. The absorbed photons are gone; they are not being intercepted. But loss creates a practical problem for the key rate: the number of photons Bob successfully detects per second. If only one photon in a thousand arrives, Alice must send a thousand pulses for every bit Bob records. The laser can fire at high rates, perhaps hundreds of millions of pulses per second, but the detectors on Bob's side have their own limitations. The result is that raw key generation rates over long fibers are measured in thousands of bits per second, not millions; for many applications, that is fine. For applications that demand high-speed bulk encryption, it is a constraint.

The wavelength of the photons matters. Glass fiber has a well-known "low-loss window" around 1,550 nanometers, the same wavelength used by long-haul telecom systems. Many QKD systems are designed to operate at or near this wavelength to minimize absorption. This choice also has a practical benefit: 1,550 nm photons can travel on the same fiber infrastructure already installed for telecommunications, sometimes sharing the glass with classical data signals on different wavelength channels. That coexistence simplifies deployment because you do not always need to lay new cable.

Temperature changes affect the fiber too. Glass expands and contracts slightly with temperature, which shifts the timing and phase of photon arrivals. Over a long outdoor fiber, temperature swings between day and night can introduce phase errors that superficially look like eavesdropping. This is one of the reasons real QKD systems constantly monitor error rates and recalibrate. The fiber is not a passive conduit; it is a dynamic environment that the system must track continuously.

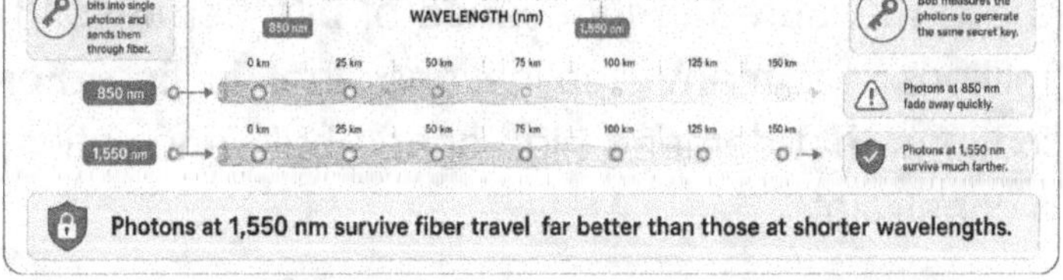

Diagram 6.4: How fiber loss and wavelength interact

Detecting Single Photons

If making a single-photon source is hard, detecting a single photon is arguably harder. Consider the scale of what you are asking. A photon at 1,550 nm carries roughly 0.8 electron-volts of energy. An electron-volt is an almost vanishingly small unit of energy; a single photon carries roughly 10 to the minus 19 joules, far less energy than a biological cell uses in a single chemical reaction. Registering that event reliably, quickly, and without false alarms is a feat of engineering precision that researchers spent decades perfecting.

The most common detector technology in deployed QKD systems is the avalanche photodiode (APD), operated in Geiger mode. Think of it like a hair trigger. The detector is held at a high voltage, just below the threshold at which it would spontaneously break down. When a photon arrives and is absorbed, it kicks an electron loose. That single electron triggers an avalanche: a cascading chain reaction of collisions that multiplies the initial charge into a detectable electrical pulse. The device is reset and ready for the next photon. This "click" is the raw data in QKD: the recorded arrival time of a qubit.

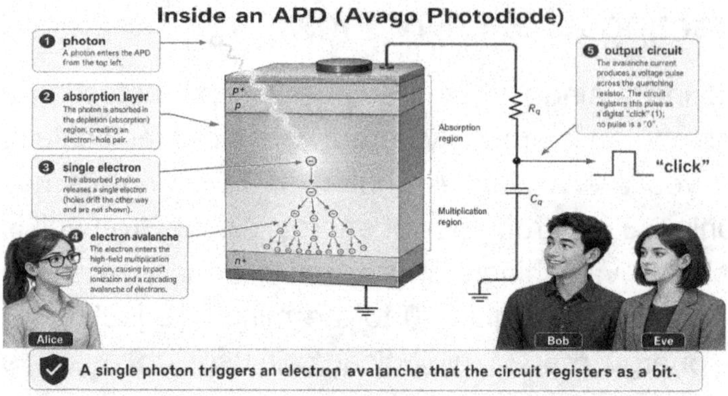

Diagram 6.5: Inside an avalanche photodiode detector

APDs have a significant weakness called dark counts. Even in complete darkness, with no photon arriving, the detector occasionally fires. Thermal noise inside the semiconductor generates a spurious electron that triggers the same cascade. Dark counts look exactly like real photon arrivals. They add noise to Bob's results, inflating what QKD systems call the quantum bit error rate, or QBER. A high QBER looks like eavesdropping. If the dark count rate is too high, the system cannot distinguish genuine noise from an attack and must either abort or generate a key with very low confidence. Cooling the APD reduces thermal noise and, therefore, the dark count rate. Many QKD systems operate their APDs at temperatures well below room temperature, sometimes using thermoelectric coolers and sometimes liquid nitrogen, depending on the performance requirements.

A more advanced detector technology is the superconducting nanowire single-photon detector, or SNSPD. A thin superconducting wire, narrower than most bacteria, is cooled to within a degree or two of absolute zero (about -269 degrees Celsius). At that temperature, it carries current with no resistance. When a photon is absorbed, it breaks superconductivity locally, causing a brief spike in resistance that registers as a click. SNSPDs offer dramatically lower dark counts, faster reset times, and better detection efficiency than APDs. Their drawback is the cooling requirement. Keeping a detector at a few Kelvin requires a specialized refrigeration unit that is large, power-hungry, and expensive. Some high-performance research systems and a growing number of commercial products use SNSPDs. As cryogenic cooling technology matures and shrinks, SNSPDs are expected to become more common in deployed systems.

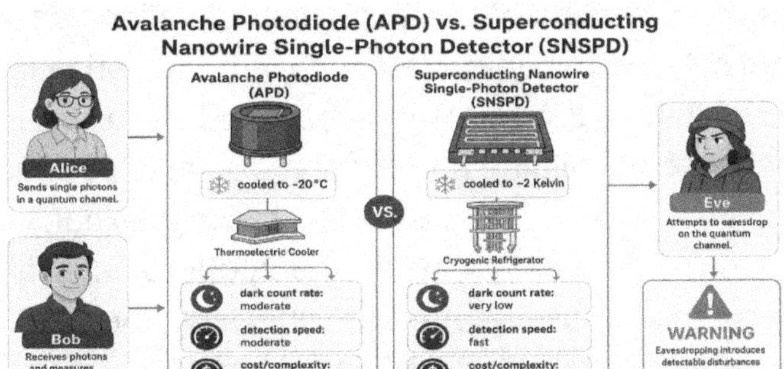

Diagram 6.6: Comparing APD and SNSPD detector types

There is another detector property worth knowing: timing jitter. When a photon arrives, the detector does not record the moment perfectly. There is a small, random delay between the photon's arrival and the electrical click the circuit records. This spread in timing is jitter, and it limits how fast the detector can accurately assign photon arrivals to clock slots. High jitter means that at very fast pulse rates, two adjacent pulses might overlap in the recorded timing data, introducing errors. Detector engineers work hard to minimize jitter, and the quest for faster, lower-jitter detectors is one of the active fronts in QKD hardware development.

Keeping the System Aligned

A QKD link is not a "set it and forget it" device. It is a precision optical instrument operating in an environment that never stops changing. Temperature, vibration, mechanical stress, and even the slow gravitational settling of buildings shift the alignment of the optical components over time. In a polarization-based system, a tiny rotation of the fiber due to twisting or bending can shift the polarization axis and introduce errors. In a phase-based system, thermal expansion changes the path length inside the interferometer and scrambles the phase reference. Either way, the system must continuously detect and correct these drifts.

The process of measuring and correcting for these effects is called calibration, and in a real QKD link it runs constantly in the background. The system monitors the QBER in real time. When the QBER rises above a threshold, the software triggers a calibration procedure, sending known reference pulses and adjusting optical elements (often tiny motorized waveplates or phase shifters) until the error rate drops back to an acceptable level. If the drift is fast, calibration can consume a significant fraction of the link's capacity. This is one of the hidden costs of real-world QKD that simple protocol descriptions never capture.

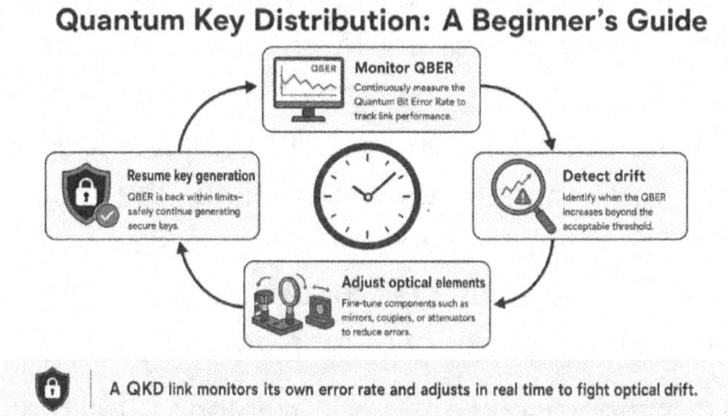

Diagram 6.7: The calibration feedback loop

Think about what temperature variation does to an outdoor fiber. In a city, a buried fiber cable might experience temperature swings of thirty degrees Celsius between winter nights and summer afternoons. Glass has a small but non-zero coefficient of thermal expansion. Over a hundred kilometers of fiber, that expansion changes the effective path length by roughly a hundred meters, which entirely changes the phase of any interferometric measurement. The system must track these changes with reference signals and compensate for them before Alice and Bob compare bases. Failure to do so would cause the system to report a catastrophic QBER and abort key generation, not because of Eve, but because of the weather.

Mechanical vibration is an equally persistent challenge. Fiber cables in data centers run alongside air conditioning units, cooling fans, and servers that vibrate at various frequencies. Each vibration slightly perturbs the polarization state or phase of photons traveling through the fiber. Short vibration bursts show up as brief QBER spikes. Long-sustained vibration from, say, heavy machinery in an adjacent room can make a link temporarily unusable. Engineers address this with vibration isolation mounts, active feedback systems, and sometimes by routing the fiber to minimize exposure to mechanical noise sources. It is unglamorous work, but it is what separates a working QKD link from a laboratory demonstration.

Timing synchronization is another calibration challenge. Alice and Bob must share a common clock so they know which of Bob's detections corresponds to which of Alice's transmitted pulses. Misaligned clocks mean mismatched bits, which looks like random noise on top of the quantum signal. Most QKD systems send a bright classical synchronization pulse alongside the quantum channel, either on a separate fiber or on a separate wavelength channel on the same fiber. This synchronization signal must be precise to within nanoseconds. It is also, as you will see in the next section, a potential attack surface.

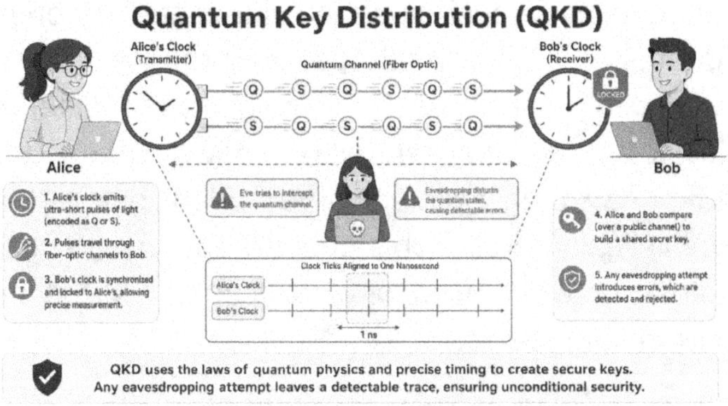

Diagram 6.8: Synchronization between Alice and Bob

When Hardware Can Be Fooled

Here is a sobering fact: several real QKD systems have been defeated, not by breaking the quantum physics, but by attacking the hardware. Researchers have demonstrated practical attacks that exploit gaps between the ideal device assumed by the security proof and the actual device sitting in the rack. These are not theoretical vulnerabilities. They have been demonstrated in published experiments against systems that were, at the time, commercially available or being prepared for deployment.

The most famous class of hardware attack targets the detectors. The core principle of QKD security assumes that Bob's detectors are fair and indifferent. When a photon arrives, the detector clicks, or it does not, based on quantum mechanics. An eavesdropper should not be able to influence whether a specific detector clicks. It turns out that real APDs are not perfectly fair. They have a property called efficiency mismatch: under certain conditions, such as when illuminated by bright pulses outside their normal operating range, different detectors in Bob's system can have different sensitivities. An attacker who knows about this mismatch can craft specially timed bright pulses that deterministically force Bob's chosen detector to click, regardless of what Alice originally sent. This is the detector blinding attack.

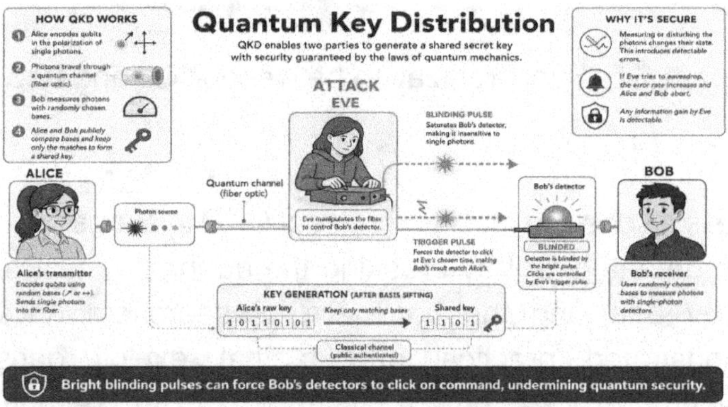

Diagram 6.9: The detector blinding attack

The detector blinding attack is elegant in a disturbing way. It works at the hardware level without ever disturbing the quantum states in the fiber. From Alice's perspective, the quantum channel looks fine. From Bob's perspective, the detectors are producing outputs, so the system appears to be working. But the outputs are under Eve's control. The QBER may even stay low, because Eve is carefully crafting the trigger pulses to match Alice's original signals. The attack compromises security without triggering any standard alarms.

Other hardware attacks target different components. Timing side-channel attacks exploit the fact that Bob's detectors may respond slightly differently to photons depending on their polarization or wavelength, leaking information about Bob's measurement basis. Trojan-horse attacks go in the other direction: Eve shines light back up the fiber toward Alice's equipment, hoping that light will reflect off Alice's internal optical components in a way that reveals her current polarization setting. Some modulators inside Alice's box have slightly different reflectivities depending on what voltage they are currently set to, and Eve can exploit that to learn Alice's basis without ever touching the quantum channel.

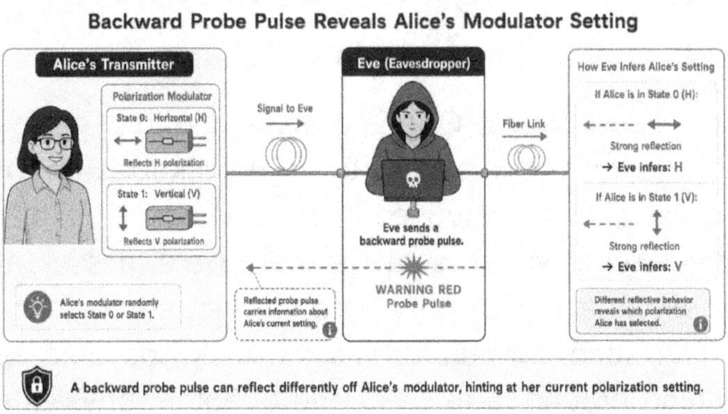

Diagram 6.10: The Trojan-horse attack on Alice's transmitter

What should you make of these attacks? Two things. First, they show that the gap between the ideal and the real is where attacks actually occur. No competent adversary tries to break quantum physics. Competent adversaries look for engineering shortcuts, component quirks, and implementation assumptions. Second, the community's response to these attacks has been thoughtful and systematic. Countermeasures include optical isolators that block backward-propagating light (defeating Trojan-horse attacks), watchdog detectors that monitor for unexpected bright pulses (detecting blinding attempts), and a rigorous discipline called device-independent QKD, or DI-QKD, which seeks to prove security without assuming anything about the internals of Alice's and Bob's devices. DI-QKD is still largely at the research stage, but it represents the most principled long-term response to the hardware attack problem.

Why Engineering, Not Just Physics, Determines Security

Imagine two houses. One is designed with a perfect lock, mathematically proven to be unpickable, on the front door. The other house has the same lock but also has reinforced windows, a monitored alarm system, and motion-sensitive lights. The first house trusts the lock. The second house trusts the lock and defends everything around it. In QKD, the quantum physics is the perfect lock. The engineering is everything around it.

The physics of QKD provides a proof of security under a specific set of assumptions. Those assumptions include: the photon source emits at most one photon per pulse; the detectors measure exactly what the protocol specifies; Eve cannot access Alice's or Bob's equipment directly; and the classical communication channel is authenticated. When all those assumptions hold, the security guarantee is real and strong. When any assumption fails, the guarantee may weaken or vanish entirely.

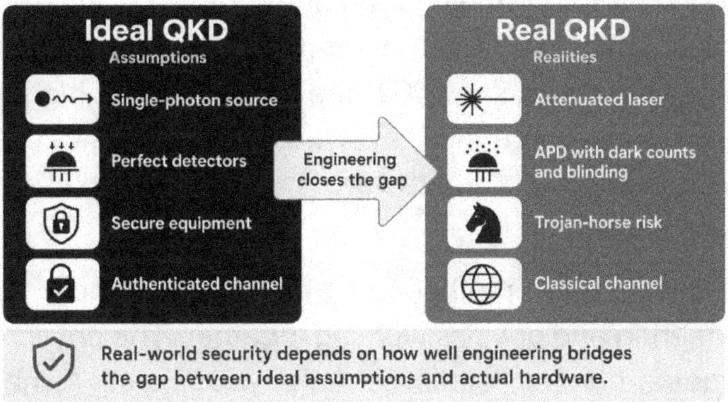

Diagram 6.11: The gap between ideal QKD and real QKD

This does not mean QKD is broken or untrustworthy. It means QKD security is a living engineering discipline. Every hardware flaw researchers find and publish is one vendors can fix in the next generation of hardware. Countermeasures are real; they work, and the community that develops and tests them is active and serious. What it does mean is that purchasing or deploying a QKD system and trusting it purely because "quantum physics guarantees security" is insufficient. You also need to ask: has this specific implementation been tested against known hardware attacks? Is the firmware up to date? Is the device operated within its certified parameters?

Certification and testing standards for QKD hardware are still maturing. Several national standards bodies and international organizations are developing frameworks that specify the tests a QKD device must pass to claim a given security level. This is analogous to how cryptographic algorithms must undergo standardization before government agencies trust them, except that, for QKD, the complexity is higher because hardware is involved. The physics is standardizable. The hardware implementations are diverse and evolving.

The decoy-state method is a perfect example of engineering and physics working together. The physics of photon-number statistics tells you exactly what pattern to expect in a world with no eavesdropping. The engineering builds a source that can switch intensity levels rapidly. The mathematics derives a provably tight bound on Eve's information from the measured statistics. The result is a security argument that is rigorous at every layer. That combination of physics, mathematics, and engineering is the model for what good QKD security looks like in practice.

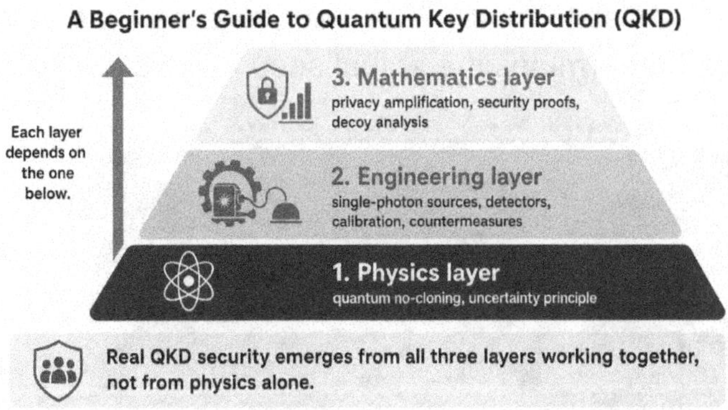

Diagram 6.12: Three layers of QKD security

A useful mental model is to think of security assurance in QKD as a three-part test. First: does the physics work correctly in this device? That means verifying the source, channel, and detectors behave as the security proof assumes. Second: has the device been hardened against known implementation attacks? That means Trojan-horse isolation, detector monitoring, and timing channel mitigation. Third: is the device operated, maintained, and updated by people who understand what they are certifying? A perfectly built QKD system operated by a team that does not understand its calibration requirements is a security risk regardless of the underlying physics.

Engineers working on today's deployed QKD networks take all three questions seriously. Some of the most important contributions to the field over the past decade have come not from new quantum physics discoveries but from security analyses of real hardware and the countermeasures developed in response. The phrase "quantum-secured" on a product box is a marketing statement. The engineering discipline behind it determines the actual security level.

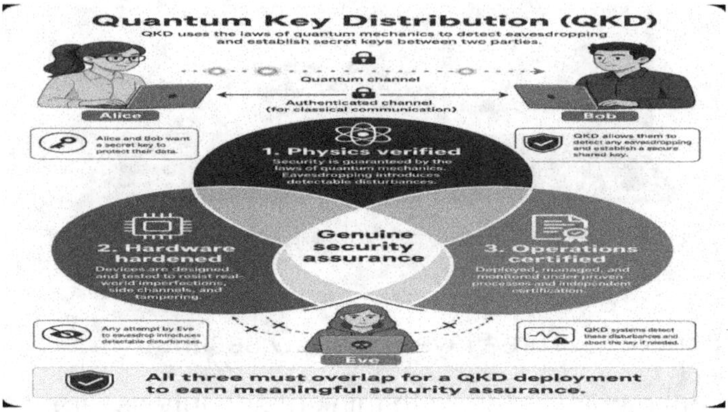

Diagram 6.13: The three-part security assurance test

The story of QKD hardware is ultimately an optimistic one. Each weakness found has led to a documented countermeasure. Each documented countermeasure has been incorporated into better hardware. The field is young enough that this cycle is still visible in the literature: an attack paper appears, a defense paper follows within a year or two, and the next generation of hardware ships with the defense built in. This is not unusual in the history of security technology. It is, in fact, how security gets built: iteratively, honestly, and with a community that rewards finding problems rather than hiding them.

Key Concepts Checklist

• I understand why a true single-photon source is hard to build, and why an attenuated laser is the practical alternative.

• I can explain, in one sentence, what decoy states are for and why they work.

• I know that fiber loss, not eavesdropping, is the main practical limit on QKD distance.

• I understand what an avalanche photodiode does, why it needs cooling, and what dark counts are.

• I know that real QKD systems must constantly recalibrate to compensate for temperature, vibration, and timing drift.

• I understand that real-world attacks on QKD systems have targeted hardware quirks, not quantum physics.

• I understand that QKD's security guarantees apply to ideal devices, and that engineering must close the gap to real devices.

Questions to Explore

• Why is a source that sometimes emits two photons a security problem rather than just a technical inefficiency? What specific action does an eavesdropper take to exploit it?

• If fiber absorbs most photons over long distances, what are the practical options for building a QKD link between two cities that are three hundred kilometers apart?

• Should you trust a QKD device because of the underlying physics, or because of how thoroughly it has been built and tested? What would a responsible purchasing conversation look like?

- What kinds of attacks become possible when the target is the hardware rather than the mathematical protocol? How is that class of attack different from attacks on classical encryption?

Try This / Reflect

Search for a short video tour of a working quantum optics laboratory or a commercial QKD system installation. As you watch, try to identify the three main stages described in this chapter: the photon source, the transmission channel (fiber or free space), and the detector module. Pay attention to how much of the physical apparatus is devoted not to sending and receiving photons but to keeping everything aligned, stable, and calibrated. Notice the vibration isolation tables, the temperature-controlled enclosures, and the feedback electronics. Reflect on how different the real machine feels from the clean block diagram you might have imagined. Then ask yourself: given everything you saw, what part of the system would you want a security auditor to check first?

If you have a background in any precision instrument, whether it is a spectrometer, a medical imaging device, or a high-precision measuring tool, try to draw parallels between the calibration challenges you know and the calibration challenges described in this chapter. The physics is different, but the engineering discipline of fighting drift, noise, and environmental interference is universal.

Glossary Terms

• **Single-photon source** - A device designed to emit exactly one photon per triggered pulse, eliminating the multi-photon vulnerability that enables the photon-number-splitting attack.

• **Decoy states** - Extra pulses of deliberately varied brightness, sent by Alice alongside signal pulses, whose arrival statistics let Alice and Bob detect selective eavesdropping and bound how much information an attacker could have gained.

• **Optical fiber** - A thin strand of extremely pure glass that guides light by total internal reflection, used in QKD to carry individual photons between Alice and Bob over distances of tens to hundreds of kilometers.

• **Fiber loss** - The progressive absorption and scattering of photons as they travel through glass, expressed in decibels per kilometer, which reduces the number of usable photons arriving at Bob and limits the practical range of a QKD link.

• **Single-photon detector** - A sensor, either an avalanche photodiode operated in Geiger mode or a superconducting nanowire device, capable of registering the arrival of a single photon as an electrical signal.

• **Calibration** - The continuous process of monitoring and adjusting a QKD system's optical alignment, timing, and phase reference to compensate for environmental changes that would otherwise inflate the error rate and degrade security.

Key Takeaways

• Real QKD hardware consists of a photon source, a transmission fiber or free-space channel, sensitive cooled detectors, and active calibration electronics working together as a system.

• True single-photon emission is difficult; most deployed systems use attenuated lasers, and the decoy-state method closes the resulting security gap by enabling statistical detection of photon-number-splitting attacks.

• Fiber loss, not eavesdropping, is the dominant practical constraint on QKD distance, cutting the usable photon count by roughly half every fifteen to twenty kilometers.

• Real hardware attacks, including detector blinding and Trojan-horse probing, have defeated actual QKD systems without touching the quantum physics, demonstrating that implementation security demands its own rigorous engineering.

• QKD's physics-based security proof applies to ideal devices; real security assurance comes from closing the gap through hardware hardening, honest certification, and disciplined operation.

7 Quantum Networks on the Ground

Overview

Quantum key distribution (QKD) began in a university laboratory and has since moved into the real world. Today, actual fiber-optic networks carry quantum-secured keys between cities, banks, hospitals, and government buildings. This chapter is a practical tour of how those ground-based networks are built, why they look the way they do, and what tradeoffs their engineers had to accept. You will learn why distance is the central problem in QKD, why it cannot be solved by borrowing the tricks that ordinary fiber-optic communication uses, and how today's networks navigate that problem with a clever but imperfect workaround called the trusted node.

The trusted node is the key idea in this chapter. Because a quantum signal cannot travel indefinitely through fiber without fading, and because the laws of quantum physics forbid simply boosting that signal, long-distance quantum networks break the journey into a chain of shorter hops. At each hop, a relay station receives the key, holds it briefly, and passes it on. This makes continental-scale quantum networking possible right now. But it also means each relay station knows the key, which introduces a real vulnerability the book treats honestly rather than glossing over. By the end of this chapter you will have a clear mental picture of how real quantum networks are structured, where the Chinese backbone network fits as the world's largest example, why cities have become QKD's natural early home, and what goes into actually building one of these systems.

Opening Scenario

Imagine you work in the cybersecurity department of a large bank. Headquarters is in one city. The primary data center is seven hundred kilometers away. Every day, those two sites need to exchange fresh encryption keys, and ordinary key exchange methods are increasingly seen as vulnerable to future quantum computers. Your team decides to build a quantum-secured link. You order dedicated fiber, hire engineers, and then run straight into the same wall that every QKD team hits: 700 kilometers of glass is too much. A single photon, sent from one city to the other, will be absorbed by the fiber long before it arrives. The quantum channel goes dark.

So you do what every real quantum network does. You break the seven hundred kilometers into four or five shorter segments. At the end of each segment, you place a relay station, a small, secure building with quantum hardware inside. The photons make it that far, generate a short key, and the station passes that key on to the next station, which generates a new key with its neighbor, and so on down the chain. By the time a key reaches the data center, it has traveled in pieces rather than as a single unbroken quantum flight. It works. Your two sites are now exchanging quantum keys every hour.

But here is the part that keeps your security team up at night. At every relay station along that chain, the key existed briefly in the clear. It was unencrypted on a computer as it was handed from one quantum link to the next. If an adversary somehow compromised one of those relay buildings, they could have read the key. This is not a theoretical risk; it is a structural feature of every long-distance quantum network operating today. Understanding why this trade-off exists, and what it costs, is what this chapter is about.

What This Chapter Will Cover

• Why distance is the enemy: how fiber loss creates a hard ceiling on single-link QKD and shapes every design decision that follows.

• Why you cannot just amplify: the quantum rule that makes classical relay tricks illegal for single photons.

• The trusted-node solution: how a chain of secure relay stations extends QKD to continental scale.

• The catch with trusted nodes: what it means for security when each relay sees the key in the clear.

• Real backbones around the world: the Chinese network and comparable efforts that prove ground-based QKD at scale is real.

• Metropolitan quantum networks: why cities are QKD's natural first home and what those deployments look like.

• How a network is actually built: fiber, relay siting, key management, and integration with ordinary encryption.

Why Distance Is the Enemy

Glass is not perfect. When a photon travels through an optical fiber, the glass gradually absorbs it. The longer the fiber, the more photons disappear before they arrive. This is not a flaw in today's manufacturing; it is a fundamental property of the material. Engineers measure this absorption as the attenuation rate, which describes how quickly signal strength falls off with distance. For a standard telecom fiber operating at a wavelength commonly used for QKD, roughly half of all photons are lost every fifteen to twenty kilometers.

That might sound like a minor nuisance, but consider what it means at longer distances. Start with one thousand photons at one end of a fiber. After twenty kilometers, about five hundred remain. After forty kilometers, about two hundred and fifty. After sixty kilometers, around one hundred and twenty-five. Keep going, and by the time you reach a hundred kilometers, you have a tiny fraction of your starting count. At two hundred kilometers, you might have just a handful of photons left, barely enough to form a key. At four hundred kilometers, statistically, most photons never arrive. The quantum channel has gone quiet.

Classical telecommunications solved this problem long ago. Fiber-optic networks for ordinary data run across entire continents because they use optical amplifiers. These devices take a fading light signal and boost it back to full strength at regular intervals. If you have ever been on a video call from one continent to another, those amplifiers are part of why it worked. They are placed every 50 to 100 kilometers along major fiber routes, making a globe-spanning cable reliable. For an engineer used to classical networks, the obvious question is: why not just do the same thing for QKD?

The answer lies in something introduced in the earlier chapters: the no-cloning theorem. A quantum amplifier would have to read the incoming photon, copy its quantum state, and then re-emit the copy. But copying an unknown quantum state is exactly what the no-cloning theorem forbids. The instant you try to read the photon's state to amplify it, you destroy the very quantum information you were trying to preserve. The photon's superposition collapses. The signal that arrives at the other end is no longer the original quantum message; it is just noise. This is not a technology gap that better hardware will close. It is a rule of physics, and every terrestrial quantum network must work around it rather than through it.

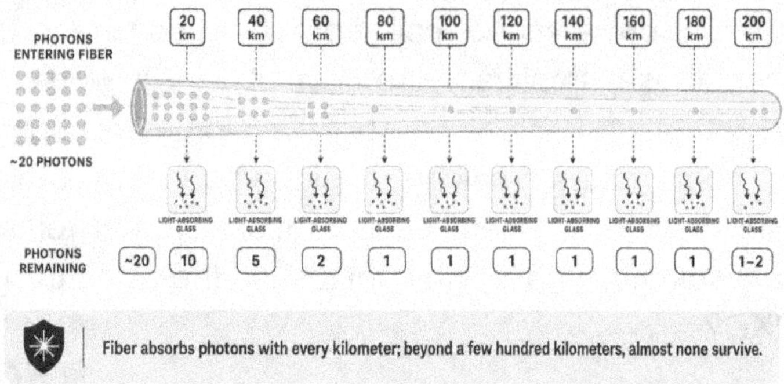

Diagram 7.1: Photon loss along a fiber link

The practical result is a distance ceiling. For most deployed QKD systems today, a single link between two nodes can sustain a usable key rate out to somewhere between one hundred and four hundred kilometers, depending on the hardware, the fiber quality, and how many secure key bits per second the application actually needs. Push that range and the raw key rate drops below a useful threshold. The quantum channel still works in principle, but the key it generates is generated so slowly that it cannot keep up with the data that needs to be encrypted.

This ceiling shapes everything about how a real quantum network is designed. Every architectural decision in the rest of this chapter is, at its root, a response to this single constraint. Engineers are not free to lay fiber from any point A to any point B and call it a quantum link. They must plan their networks around the physics of loss, choosing relay station locations so that no individual hop stretches past the survivable range. The art of building a quantum network is largely the art of working within this ceiling.

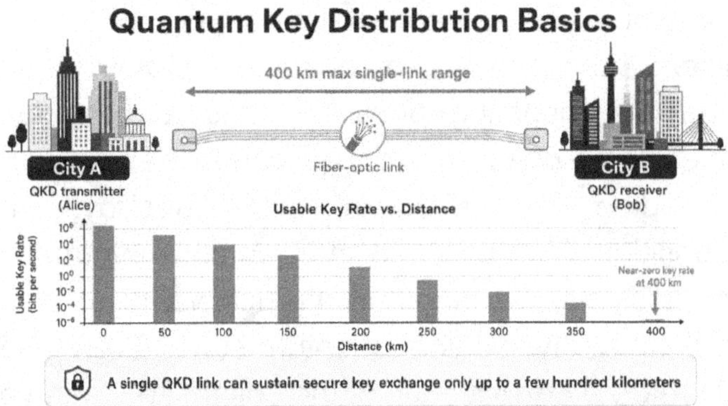

Diagram 7.2: The single-link distance ceiling

Why You Cannot Just Amplify

The no-cloning theorem is short to state but far-reaching in its consequences. It says: you cannot make a perfect copy of an unknown quantum state. That sounds abstract, so consider what it means in practice for a photon carrying quantum key information.

When a photon leaves a QKD transmitter, its polarization or phase is set to a specific quantum value. That value is the information. But here is the thing: until a measurement is made, the photon exists in a superposition. It is not secretly one thing waiting to be revealed; it genuinely has no definite value yet. The value it ends up with depends on how and when it is measured. This is the very property that makes QKD eavesdrop-evident. If an eavesdropper intercepts the photon and measures it, she forces it into a definite state, disturbing the quantum randomness and leaving detectable footprints in the key.

Now imagine you are an engineer trying to build an amplifier for this photon. To boost the signal, you need to read the incoming photon's state and produce multiple copies of it so the copies can be sent onward. But reading the photon's state destroys the superposition. You collapse the quantum information into a classical value, and that classical value is not the same thing as the original quantum state. Even if you could create a perfect copy of the collapsed state, you would be sending on a photon that has already been measured. The downstream receiver would observe a photon that behaves classically rather than quantum-mechanically. The security guarantee would evaporate.

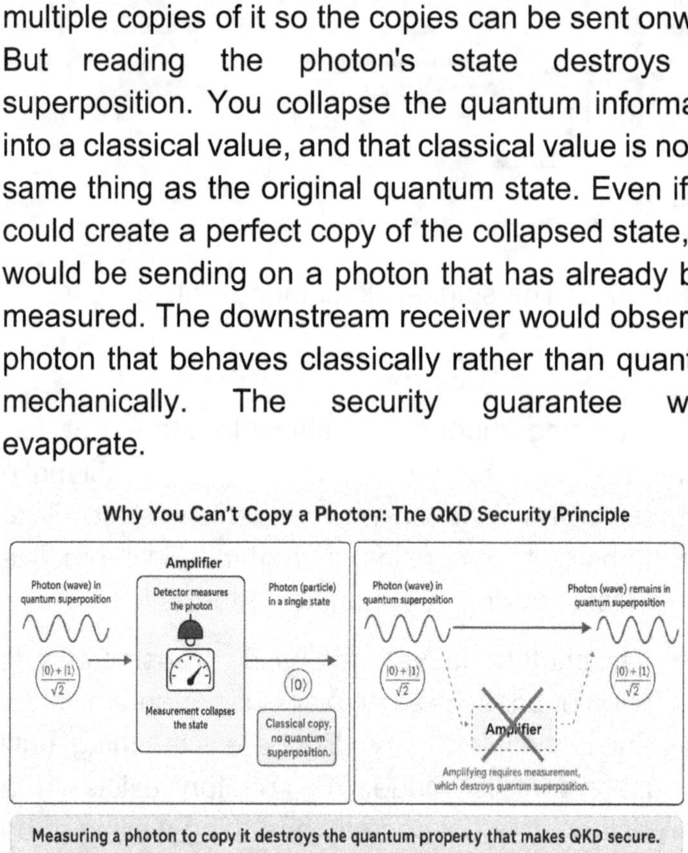

Diagram 7.3: Why amplification breaks quantum information

This is worth sitting with for a moment, because it feels counterintuitive. In everyday life, copying is easy. You copy a file, a photograph, a document. Classical information can be duplicated without limit. But quantum information lives by different rules. The superposition that carries the key information is not a value written somewhere waiting to be read off; it is a physical property that only exists until you look at it. Looking at it is the same as destroying it. There is no way to make a transparent copy.

Some readers wonder whether future technology might find a loophole. Perhaps a more sophisticated device could somehow copy a quantum state without measuring it? Physicists have studied this question carefully. The no-cloning theorem is not an engineering limitation. It follows directly from the mathematics of quantum mechanics, specifically the linearity of quantum evolution. Any process that produced a perfect copy of an unknown quantum state would violate linearity, contradicting a foundational assumption of the entire theory. Unless quantum mechanics itself turns out to be wrong, perfect quantum amplification is not coming.

This is important because it means QKD networks cannot follow the path of classical fiber-optic networks. Classical networks scaled from city to city to continent to continent primarily because amplifiers made distance a solved problem. QKD cannot use that solution. It needs a different one, and the one the field has settled on, at least for now, is the trusted node.

Diagram 7.4: Classical amplifier versus the quantum no-go

The Trusted-Node Solution

The trusted node is the engineering community's current answer to a problem physics refuses to let you solve directly. The idea is straightforward: if you cannot relay a photon without destroying it, relay the key instead. Break the long journey into segments short enough for individual QKD links to handle. At the end of each segment, place a station. That station runs one QKD session with the node behind it and another QKD session with the node ahead. It receives a key from the left, and it sends a new key to the right, using a mathematical operation to tie the two together so that the overall secret passes through.

Think of it like a bucket brigade. You want to move water from a well to a fire. The well is too far away to carry a bucket the whole distance. So you line up a chain of people. The first person fills a bucket and passes it to the second. The second passes it to the third. By the time the water reaches the fire, no single person carried it the whole way. Each person in the chain handled the bucket briefly. The water arrived. The trusted-node chain works the same way, with each station playing the role of one person in the brigade.

The specific mechanism works like this. Say you have three nodes: A, B, and C, and you want to share a secret key between A and C. A and B first run a QKD session and end up with a shared secret key they will call K1. Then B and C run their own QKD session and end up sharing K2. Now B takes K1, applies it to K2 using a simple combining operation, and sends the result to C over a classical channel. C uses K2 to reverse the combining operation, and ends up with K1. At this point, A and C both hold K1, and they can use it as their shared secret. B facilitated the handoff, but B no longer holds K1 in a usable form once the combining operation is complete.

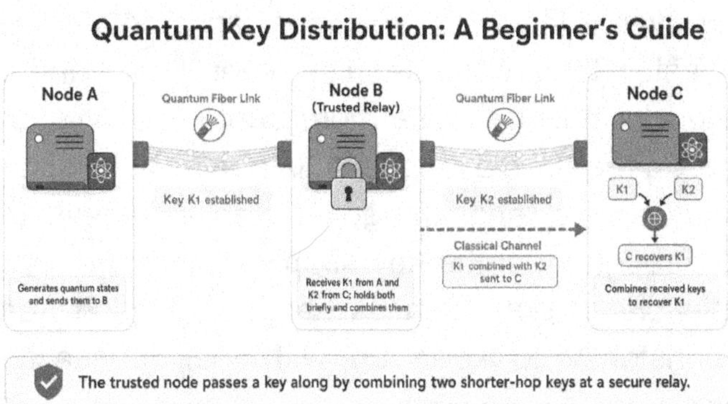

Diagram 7.5: The trusted-node bucket brigade

This relay approach extends the range of QKD to any distance, at least in principle. You can build a chain with as many nodes as you need. Each hop is short enough for photons to survive the journey. The chain can stretch across a country or a continent. The only constraint is that you must keep building secure relay stations.

The trusted-node architecture also offers a side benefit: it allows the network to use different QKD protocols, hardware, or fiber types across different segments. Node B does not care which QKD system A and B used to establish K1, as long as K1 is secure; similarly, for the B-to-C hop. Each segment is technically independent. This is useful in practice because QKD hardware is still evolving rapidly, and a national network built over several years may mix equipment generations.

Network operators have discovered another practical advantage. Because each hop is short, key generation rates on each link tend to be higher. A QKD system running over a fifty-kilometer fiber can produce keys much faster than one straining over four hundred kilometers, simply because more photons survive the journey. So breaking a long route into relay-linked short hops not only extends range; it also increases the overall throughput of the key supply. This is important for real-world users, such as banks that need fresh keys constantly to keep up with their transaction volume.

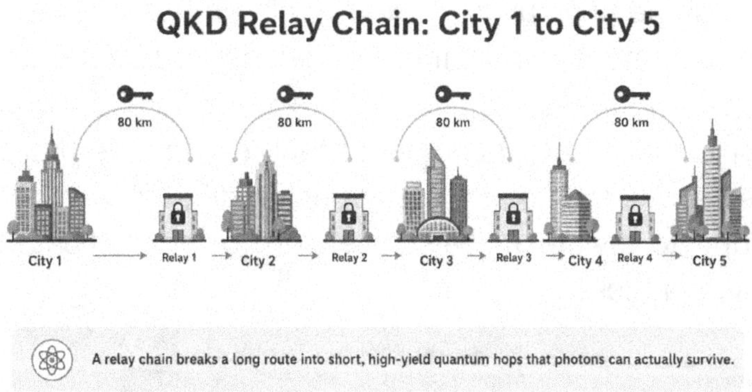

Diagram 7.6: Multi-hop key relay chain across cities

The Catch With Trusted Nodes

The trusted-node solution works. Real keys travel real distances today because of it. But every honest explanation of QKD must say plainly what this solution costs, and this book is committed to honesty.

Each relay station, each trusted node, is a place where the key exists in the clear. At the moment when Node B has received K1 from Node A and is about to generate K2 with Node C, the key material is inside Node B's hardware as ordinary, unencrypted data. It is not inside a quantum channel where eavesdropping would be detectable. It is just a number, sitting in a computer. If an adversary could compromise that computer, they would have the key.

The security of a trusted-node network is therefore not purely physics-based. It is a combination of physics and classical security measures. The quantum link between adjacent nodes is genuinely quantum-secure; eavesdropping on the fiber between them is detectable by the same principles you learned in earlier chapters. But the relay stations themselves are protected by physical locks, guards, cameras, access controls, and the operational integrity of those who run them. That is classical security: good, but not unconditional, and not immune to insider threats, coercion, or sophisticated physical attacks.

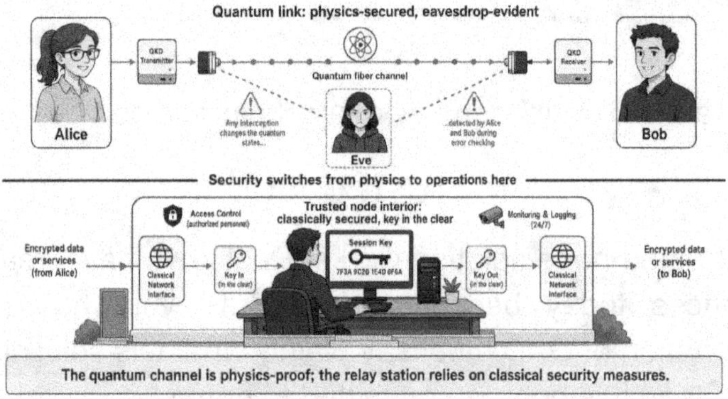

Diagram 7.7: The trusted-node security split

This catch is real and worth careful thought. Consider a national quantum network operated by a government or a major institution. The relay stations along its backbone must each be physically secured around the clock. That means buildings with reinforced walls, access restrictions, security personnel, and continuous monitoring. It means background checks for everyone who enters. It means regular audits to verify that the hardware has not been tampered with. This is not impossible; high-security facilities operate all over the world. But it adds cost, complexity, and a dependency on human trustworthiness that pure QKD theory does not require.

It also changes the threat model. In a network protected purely by quantum physics, there is no meaningful attack other than breaking the laws of physics. In a trusted-node network, an adversary with enough patience or resources can potentially focus on the relay stations rather than the quantum channel. They do not need to intercept photons. They need to get inside a building. History shows that determined adversaries are creative about getting inside buildings.

This is not a reason to dismiss trusted nodes. They are the best available solution today, and they enable quantum-secured communication at scales that would otherwise be completely impossible. But the field is clear-eyed about the limitation. A great deal of research energy is currently directed at quantum repeaters. These devices could relay a quantum signal without ever measuring it, using a phenomenon called quantum entanglement to pass the state along while leaving it unmeasured. Quantum repeaters are the subject of a later chapter. When they mature, they will remove the trusted-node catch entirely, making long-distance QKD as unconditionally secure as a single short link. Until then, trusted nodes are the honest, practical answer.

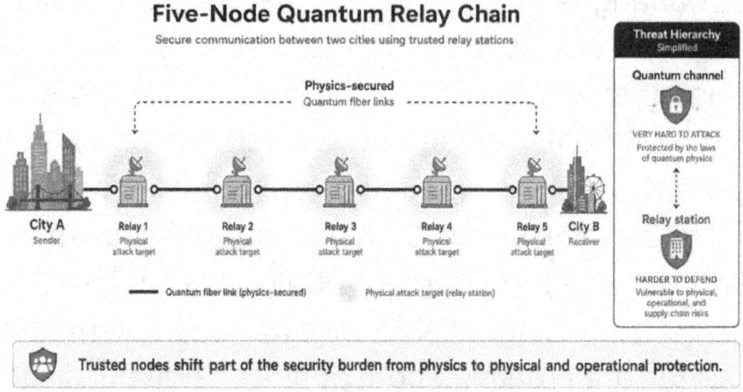

Diagram 7.8: Attack surface of a trusted-node chain

Real Backbones Around the World

The largest ground-based quantum network ever built spans more than 2,000 kilometers across eastern China, connecting Beijing in the north to Shanghai in the south and passing through several major cities along the way. It was built over roughly a decade, uses thousands of kilometers of fiber, incorporates dozens of relay stations, and is actively used by financial institutions, government agencies, and other organizations for live key distribution. This network is the clearest proof that continental-scale QKD is not a futuristic concept; it is an operating reality.

What makes the Chinese backbone remarkable is not just its length but its architecture. The relay stations along the route are not evenly spaced academic test points. They are sited based on real geography, population density, and the locations of actual users. The cities it connects are among the most economically active in the world, and the financial sector there was among the first to adopt quantum-secured key distribution for interbank communications. The network has carried millions of secure keys in routine operations, making it an enormous data point on what terrestrial QKD looks like at scale.

Other countries have built or are building comparable systems. European programs have linked laboratories across multiple countries for research purposes, with plans to extend those links into operational infrastructure. In the United States, research networks have connected universities and government facilities, demonstrating multi-node key distribution over metropolitan and regional distances. Japan and South Korea have both demonstrated QKD links between financial institutions and government sites. The common thread across all of these efforts is the same basic architecture: fiber, relay stations, trusted nodes, and integration with classical encryption infrastructure.

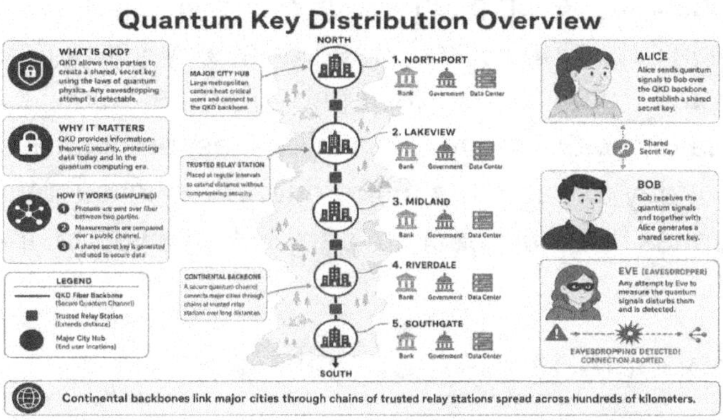

Diagram 7.9: Continental backbone layout concept

It is worth pausing to acknowledge something that the existence of these networks tells us. Building a quantum backbone requires sustained political will, substantial capital investment, and years of engineering effort. The Chinese network required a major national program to pull off. The fact that governments and large institutions have chosen to make those investments tells you something about how seriously they take the long-term threat from quantum computers to current encryption. They are not waiting for quantum computers to arrive before starting to protect their most sensitive communications. They are building the infrastructure now, while there is still time to deploy and operate it before the threat matures.

This is sometimes called quantum readiness or harvest-now-decrypt-later preparedness: the idea that adversaries may already be storing encrypted data today, intending to decrypt it once a sufficiently powerful quantum computer exists. A government agency sending a highly classified communication over conventional encryption today might see that communication decrypted in ten or twenty years if the encrypted data was captured and stored. For information with a long sensitivity lifespan, that is a present-day threat, not a future one. National quantum backbones are, at least in part, an answer to that threat.

Metropolitan Quantum Networks

If long-distance quantum networking is the hard problem, metropolitan QKD is the easier one. Most cities are compact enough that key distances between locations remain well within the range that a single

QKD link can handle without any relay stations. A city where the farthest two points are forty or fifty kilometers apart is almost perfect terrain for QKD. Photons survive the journey. No relay is needed. The quantum link between, say, a hospital records system and a health authority data center can be a single, uninterrupted quantum channel with no trusted nodes and no classical security compromises in the middle.

This is why cities have become QKD's natural early home. The physics is friendlier, the economics are easier because you need less equipment per route, and the use cases are highly concentrated. A financial district where dozens of banks cluster within a few kilometers of each other can be served by a metropolitan QKD ring, a loop of quantum fiber with branches to each member institution, allowing any two of them to exchange keys without any point in between ever holding the key in the clear.

Metropolitan networks also tend to have more fiber infrastructure to work with. Cities have extensive underground ducting from decades of telecommunications investment. Dark fiber, meaning fiber strands that were laid in the ground but are not currently carrying any traffic, is often available for lease. QKD operators in several cities have taken advantage of this existing infrastructure, leasing dark fiber and deploying their quantum hardware on top of it rather than digging new conduits. This significantly reduces deployment cost and time compared to a greenfield installation.

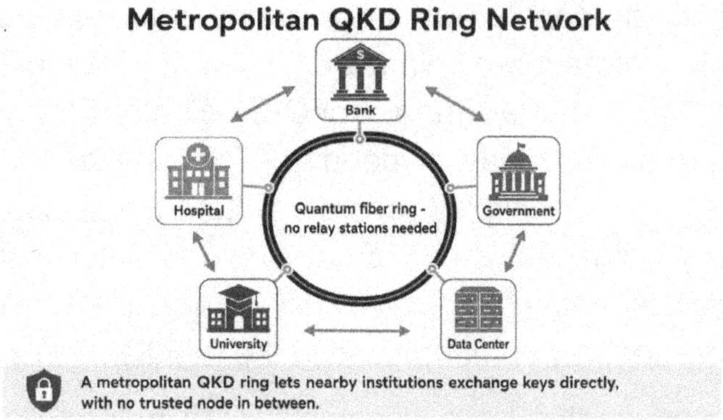

Diagram 7.10: Metropolitan QKD ring connecting institutions

The use cases in metropolitan deployments are highly concrete. Banks use metropolitan QKD to exchange keys between their branches and central processing sites, securing interbank settlement traffic. Hospitals and health systems use it to transmit patient records between care facilities and central data repositories, where regulatory requirements for long-term data protection are particularly stringent. Some government cities have deployed QKD to connect legislative, judicial, and executive facilities, ensuring that classified communications within the city are quantum-secured end-to-end.

Another important feature of metropolitan networks is their ability to be designed for high redundancy. A ring topology, where the fiber loops back on itself, means that if any single fiber section fails, the traffic can route around the break in the opposite direction. Adding a second ring, or connecting multiple rings, gives the network multiple paths between any two nodes.

Classical telecom engineers have used ring topologies for exactly this reason for decades. QKD metropolitan networks can inherit the same resilience principles, since the quantum hardware operates over fiber and does not require fundamental changes to how fiber rings are planned.

The density of users in a city also makes it easier to justify the capital cost of QKD infrastructure. Spreading the equipment cost across dozens of users, each paying for access to the quantum-secured key distribution service, yields a more favorable economic calculation than a long rural backbone that serves only a handful of endpoints. Some observers expect that metropolitan QKD networks, rather than long-distance backbones, will be the first commercial deployments to achieve self-sustaining financial viability without government funding.

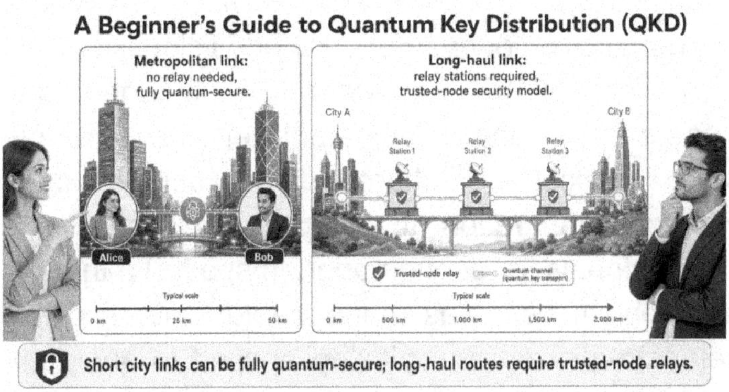

Diagram 7.11: Single-link metropolitan QKD versus multi-hop backbone

How a Network Is Actually Built

Understanding the architecture of a quantum network is one thing. Appreciating what it actually takes to build one is another, and the practical reality is considerably more involved than any diagram shows.

The first decision in any QKD deployment is whether to use dedicated fiber or to share existing fiber with classical traffic. Dedicated fiber is cleaner. The quantum channel does not have to compete with the intense classical light signals that would otherwise be traveling through the same strand. Bright classical light can scatter into the quantum channel and create noise that obscures the single-photon signal, a phenomenon called Raman scattering. For this reason, the safest choice is to give the QKD system its own dedicated fiber strand, isolated from all classical traffic.

But dedicated fiber is expensive. Buying or leasing a fiber strand solely for quantum use adds significantly to the cost of a QKD deployment. Researchers and engineers have spent considerable effort figuring out how to coexist on shared fiber, using careful wavelength separation to keep the classical and quantum signals far enough apart in spectrum that the noise problem becomes manageable. This is called wavelength-division multiplexing applied to quantum-classical coexistence. It works, with caveats, and it dramatically reduces the cost of deploying QKD on existing infrastructure. Several metropolitan networks operate this way.

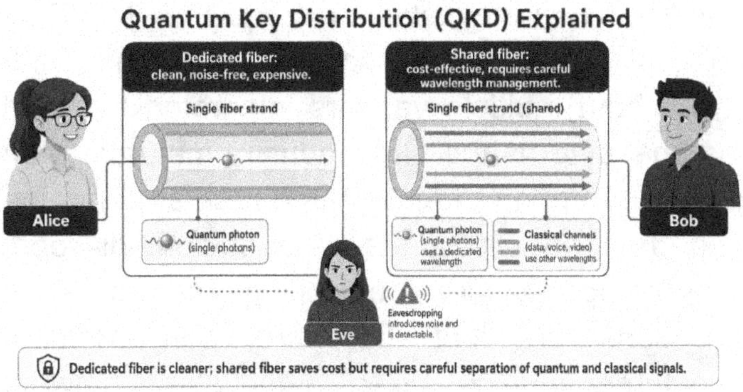

Diagram 7.12: Dedicated fiber versus shared fiber for QKD

Once the fiber decision is made, the next challenge is siting the relay stations for any links that exceed the single-hop range. Relay stations need several things. They need physical security, which means, at minimum, a locked room with access logging, and ideally a proper secure facility with environmental controls to keep the quantum hardware within its operating temperature range. They need both primary and backup power because QKD hardware that loses power loses the ability to generate keys and may lose any keys it is holding in transit. They need network connectivity for the classical channel that carries post-processing and key-management traffic, which runs alongside the quantum channel. And they need maintenance access, since the optical hardware inside the relay requires periodic alignment and calibration.

None of this is exotic. Telecommunications companies operate equipment rooms and relay stations continuously. But quantum relay stations are still less standardized than telecom equipment. Each deployment tends to be somewhat custom, with hardware from multiple vendors integrated by hand. This is changing as the QKD industry matures. More vendors are offering pre-integrated rack systems that can be dropped into a standard equipment room. Standards bodies are developing interoperability specifications so that hardware from different manufacturers can be mixed in the same network. Progress is real, but anyone building a QKD network today should expect more integration work than a comparable classical network deployment would require.

Key management is the operational challenge that often surprises newcomers to QKD. The network generates keys continuously. Those keys must be stored securely until they are used, delivered to the applications that need them, and destroyed after use to prevent reuse. In a large metropolitan network with dozens of member institutions, the key management layer becomes a substantial software system in its own right. It needs to track which keys are available, deliver them to the right applications at the right time, handle cases where a QKD link goes down or the key supply runs low, and log all key usage for audit purposes.

Some organizations use a key management server that sits between the QKD hardware and the applications. The quantum network delivers keys to the key management server, and applications request keys from the server as needed. The server handles the bookkeeping: which keys are fresh, which have been used, which links are currently generating new supply, and how to route a key request if the most direct quantum path is unavailable. This abstraction layer also enables mixing QKD-derived keys with keys generated by post-quantum cryptographic algorithms, a hybrid approach that several security standards recommend for the transitional period we are currently in.

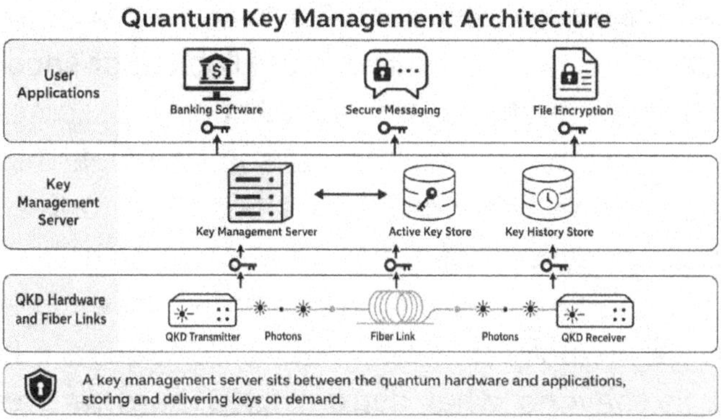

Diagram 7.13: Key management layer in a QKD network

Integration with classical encryption is another practical reality. A QKD network does not replace all the other security infrastructure an organization has. It supplements it. The quantum keys generated by a QKD link are typically fed into standard symmetric-key encryption algorithms, such as AES (Advanced Encryption Standard), the workhorse of data encryption worldwide. The QKD link ensures that the AES key was distributed in a quantum-secure manner. AES itself does the actual work of encrypting the data stream. This means organizations do not have to throw out their existing encryption infrastructure to adopt QKD. They add a quantum-secured key supply layer on top of what they already have.

This integration model also helps with one of the practical challenges of QKD: key generation rate. A QKD link over fiber produces keys at a certain rate, measured in bits per second. That rate depends on the distance, the hardware quality, and the noise level. For a fifty-kilometer link with good hardware, a typical rate might be in the tens to hundreds of kilobits per second for secure key material. For a two-hundred-kilometer link, the rate drops. For comparison, a single high-definition video stream requires around ten to twenty megabits per second of bandwidth. Encrypting all that data with one-time pad encryption, in which each key bit encrypts exactly one data bit, would require a key supply far beyond what current QKD hardware can deliver.

The standard solution is to use the quantum key to protect a much shorter AES session key. AES operates securely with a one-hundred-twenty-eight- or two-hundred-fifty-six-bit key. A QKD link can generate several of those keys per second even at moderate distances. The session key refreshes frequently, say every few minutes or hours, keeping the window of exposure small even if a key were somehow compromised. This hybrid architecture captures the security benefits of quantum key distribution for the key exchange step while using classical encryption to handle the heavy lifting of data encryption.

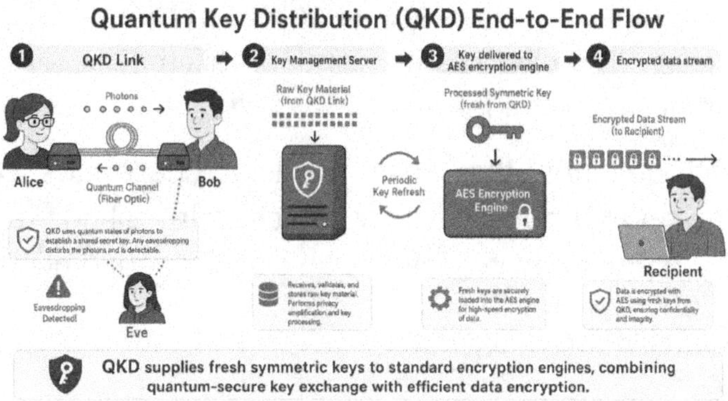

Diagram 7.14: QKD integration with classical encryption

Key Concepts Checklist

• I understand why fiber absorption creates a hard distance ceiling for single-link QKD, making long-distance quantum networking impossible without relay stations.

• I can explain why the no-cloning theorem prevents classical amplifiers from solving the distance problem for quantum signals.

• I can describe what a trusted node does: it receives a key from one QKD link and passes it to the next, briefly holding the key in the clear.

• I understand the core security catch of trusted nodes: each relay station holds the key unencrypted for a moment and must therefore be physically and operationally secured.

- I can identify at least one real-world terrestrial quantum network and describe its general architecture.
- I understand why cities are friendlier territory for QKD than long intercity routes, and what metropolitan network topologies look like.
- I can describe the main practical challenges of building a QKD network: fiber choice, relay siting, key management, and integration with classical encryption.

Questions to Explore

- If every trusted node in a national quantum network must be physically secured to a high standard, what does that requirement say about which organizations are capable of operating such a network? Could a private company do it, or does this naturally push toward government control?
- A country might choose to build and operate a quantum backbone even though its trusted nodes introduce a vulnerability that pure QKD theory does not have. What strategic or political reasons might make that trade-off worthwhile?
- Cities have been described as the natural early home of QKD because distances are short and users are clustered. Can you think of other environments, outside cities, that might share those characteristics? What kinds of facilities or campuses might be good candidates for QKD without any relay nodes?
- The trusted-node model shifts part of the security guarantee from physics to operations and human trust.

How does this compare to the promise you have heard throughout this book that the laws of physics secure QKD? Is there a meaningful way to describe a trusted-node network as still being "quantum-secured"?

Try This / Reflect

Look up a map or schematic of a real quantum communication backbone. Maps have been published in research papers and news articles covering the Chinese Beijing-to-Shanghai network, and similar diagrams exist for European and Asian QKD research networks. As you trace the route, identify each relay station or node along the way. For each one, remind yourself: at this point in the chain, the key was briefly in the clear. Ask yourself how far apart the relay stations are, what sorts of buildings they might be located in, and what it would take to secure each one.

Then, on a piece of paper, sketch your own imaginary three-city quantum network. Pick three cities that are roughly two hundred to three hundred kilometers apart. Mark where you would place your relay stations based on the distance constraints you learned in this chapter. Label the quantum links between nodes. Then label the trusted nodes and write next to each one what classical security measures you would apply. This exercise turns the abstract architecture into something concrete and personal. You may find that the practical complexity of what you are drawing gives you a deeper appreciation for the engineering effort behind the real networks that already exist.

Glossary Terms

• **Attenuation** - The gradual loss of signal strength as a photon travels through fiber, caused by the glass absorbing or scattering light over distance.

• **Trusted node** - A secure relay station in a quantum network that receives a key from one QKD link and passes it to the next, holding the key briefly in unencrypted form.

• **No-cloning theorem** - The quantum physics rule stating that it is impossible to make a perfect copy of an unknown quantum state, which prevents photon amplification.

• **Metropolitan QKD network** - A quantum key distribution network operating within a city or urban area, where distances are short enough that relay stations are often unnecessary.

• **Key management** - The set of processes for storing, delivering, refreshing, and destroying encryption keys in a secure and organized way across a network.

• **Dark fiber** - Fiber-optic cable that has been installed in the ground but is not currently carrying any active traffic, often available for lease by QKD operators.

• **Wavelength-division multiplexing** - A technique for combining multiple signals on a single fiber strand by assigning each signal a different wavelength of light, used to allow QKD and classical signals to share the same fiber.

Key Takeaways

• Fiber absorption limits a single QKD link to a few hundred kilometers, and this ceiling shapes every aspect of quantum network design.

• The no-cloning theorem makes quantum amplification physically impossible, so QKD networks cannot use the same relay tricks as classical fiber-optic infrastructure.

• Today's long-distance quantum networks use trusted-node relay chains, where secure stations pass keys hop by hop along a route.

• Each trusted node holds the key in the clear for a moment, meaning a trusted-node network's security depends partly on classical physical and operational security, not only on quantum physics.

• Real quantum backbones spanning thousands of kilometers already operate, with the Chinese Beijing-to-Shanghai network as the largest example, while metropolitan networks in cities worldwide are proving that QKD at shorter ranges is technically and commercially practical.

8 Quantum Keys from Space

Overview

If you tried to send a quantum key across a continent using fiber, you would lose nearly every photon long before the message arrived. Fiber is glass, and glass is imperfect. Every kilometer absorbs a little of the signal. After a few hundred kilometers, the signal is so weak that quantum key distribution (QKD) grinds to a halt. The workaround explored in earlier chapters, the trusted-node relay, gets the key moving again but only by accepting a security compromise at each relay point. A completely different solution avoids that compromise strikingly: it routes the photons through space.

The vacuum of space absorbs almost nothing. A photon beamed upward through a thin slice of atmosphere and then across hundreds or thousands of kilometers of near-vacuum loses far less energy than a photon fighting its way through kilometers of dense glass. A satellite in low orbit can therefore act as a bridge, receiving photons from one ground station and handing keys along to another station thousands of kilometers away. This chapter follows the logic of that idea from first principles, through working demonstrations, to honest practical limits. You will learn why space is so photon-friendly, how the satellite relay works, what the landmark experiments actually showed, and what challenges the technology must still overcome before satellite QKD becomes a routine service.

Opening Scenario

Picture two research centers, one near the Atlantic coast and one far inland on the other side of a mountain range. They share sensitive data every night, encrypted with fresh keys. A fiber cable between them would be thousands of kilometers long. Even with the best amplifiers and repeaters, a quantum signal would not survive the journey intact. The centers have considered a chain of trusted relay nodes, but each node is a potential weak point. Someone would have to guard every relay site, every night.

Then a colleague suggests something that sounds like science fiction at first: aim a telescope at the sky. A satellite passing overhead can carry photons between the two sites without the signal ever touching fiber. For the few minutes the spacecraft is above the horizon, both telescopes lock on to it, photons flow, and a shared secret key accumulates. When the satellite disappears over the edge of the world, the key exchange is done. The data sent that night will be encrypted with a key that never traveled through a single meter of cable.

That scenario is not fiction anymore. Experimental work has demonstrated exactly this kind of exchange over distances that would be impossible with fiber alone. The technology is young, the pass windows are short, and there are real constraints to wrestle with, but the physics is sound, and the results are in. This chapter explains how it all works.

What This Chapter Will Cover

• Why the near-vacuum of space lets photons travel farther with less loss than fiber, and why the thin atmosphere above a ground station is the bottleneck, not the long space crossing

• How a satellite acts as a relay between two distant ground stations, either forwarding a quantum signal in real time or sharing separate keys with each station and then combining them on the ground

• The landmark demonstrations that proved satellite QKD works at planetary scale, described as illustrative milestones

• The pointing and timing challenge: how engineers aim a beam of single photons from a moving spacecraft onto a telescope moving with the rotating Earth

• The practical limits of weather, daylight, and brief pass windows that shape how satellite QKD is used today

• Whether the satellite itself should be trusted, and the research aimed at removing that assumption

• The wave of constellation programs now being planned and built to turn occasional demonstrations into networked, routine service

Why Space Beats Fiber Over Long Distances

Imagine filling a swimming pool with very clear water and then trying to shine a flashlight through it end to end. Even the clearest water scatters and absorbs light. Now imagine doing the same experiment across an equal distance of open air on a dry, cloudless night. The flashlight beam travels much farther before fading. Replace the open air with the near-vacuum of space, where almost nothing gets in the way, and the beam can travel extraordinary distances with only a tiny fraction of the loss.

Fiber optic cable is made of the purest glass ever manufactured. Engineers have spent decades refining it, and it is astonishingly clear compared with ordinary glass. But it is still glass. Every kilometer introduces a little absorption, a little scattering. For ordinary laser signals, this is manageable because you can amplify the signal periodically to restore it. With quantum signals, you cannot do that. The no-cloning theorem, which you encountered in earlier chapters, means you cannot copy a quantum state and then boost it. The signal either arrives or it does not.

The result is a hard distance ceiling for fiber-based QKD. Depending on the quality of the fiber and the detectors, that ceiling is on the order of a few hundred kilometers under practical conditions. Beyond that, the photon loss rate is too high to establish a key at a useful rate. You can push further with advanced techniques, but the improvement is incremental. If you want to span a continent or an ocean, fiber alone cannot do the job without trusted relay nodes at every intermediate hop.

Space offers a different geometry. A satellite in low Earth orbit sits roughly 500 to 600 kilometers above the surface. When a ground station beams photons upward to that satellite, the signal must pass through the thin atmosphere. Most of the absorption and turbulence is concentrated in the first 10 or 20 kilometers above the ground. Above that, the air thins quickly. The rest of the 500-kilometer journey happens through near-vacuum, which is almost perfectly transparent to the photon wavelengths used in QKD.

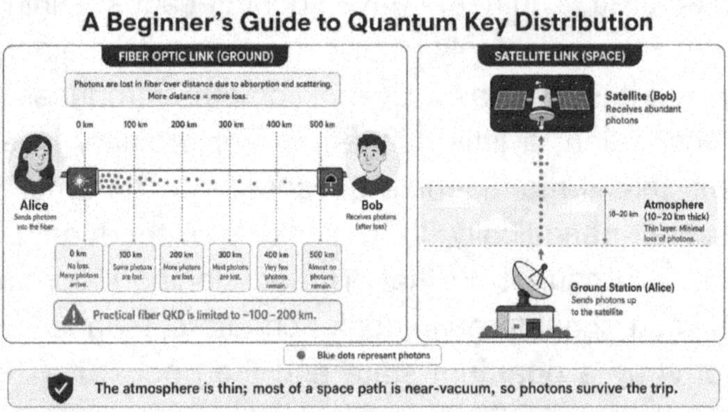

Diagram 8.1: Space vs. fiber photon loss

The total photon loss on a vertical space link is much lower than on a horizontal fiber link of comparable end-to-end distance. Researchers who have measured this directly find that a well-designed space link can deliver photons at rates that would require impossibly long fiber to match. That is the fundamental physical reason why the satellite idea works. It is not a trick or a workaround; it is a consequence of where the losses actually are. The atmosphere is the bottleneck, and it's short.

There is another subtlety worth noting. When photons travel through the atmosphere, they encounter turbulence. Packets of warm and cool air bend the beam slightly, just as heat shimmer distorts a distant object on a hot road. This turbulence spreads the beam and can cause fluctuations in the photon count at the receiver. Engineers compensate with a technique called adaptive optics, borrowed from astronomy, which uses a deformable mirror that adjusts its shape many times per second to counteract the distortion. Ground stations equipped with adaptive optics can collect significantly more photons from the same satellite pass than stations without them. The atmosphere is not a perfect window, but it is manageable.

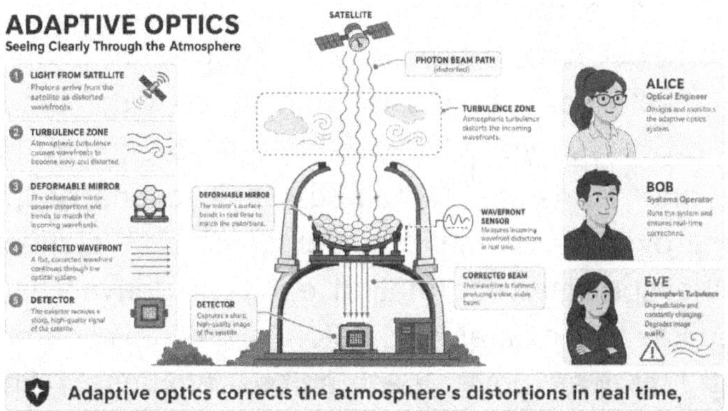

Diagram 8.2: Atmospheric turbulence and adaptive optics

Fiber QKD and space QKD are therefore not competitors in the same arena. They are tools suited to different distance regimes. Fiber works well for city-to-city links of a few hundred kilometers. Space works for continent-to-continent or country-to-country links where no practical fiber route could carry a quantum signal unaided. The two technologies complement each other in a future quantum network, with ground fiber handling local distribution and satellites bridging the wide gaps between regional hubs.

The Satellite as a Relay

Knowing that space is photon-friendly does not yet explain how a satellite actually delivers a key from one ground station to another. There are two conceptually different approaches, and it is worth understanding both because they have different security properties.

The first approach is called a downlink or uplink relay. In a downlink scheme, the satellite generates quantum signals and sends them to one ground station, then to another. Each station ends up with a secret key shared with the satellite. The satellite then securely combines those two keys, typically by using one to encrypt the other, and sends the combined result to one of the stations over a classical channel. Once both stations have that combined information, each can derive a final shared key that neither held before. They now share a secret that was never sent across any direct link between them. The satellite served as the connecting bridge.

The second approach is called entanglement-based distribution. In this scheme, the satellite generates pairs of entangled photons and sends one photon from each pair to each ground station. Because the two photons are entangled, measurements at each station are correlated, allowing the stations to derive a shared key directly. This approach has the advantage that the satellite need not know the key. It is more like a postmaster who delivers sealed envelopes than one who reads the letters before passing them on. Entanglement-based distribution is technically harder to achieve, but it offers a stronger security foundation.

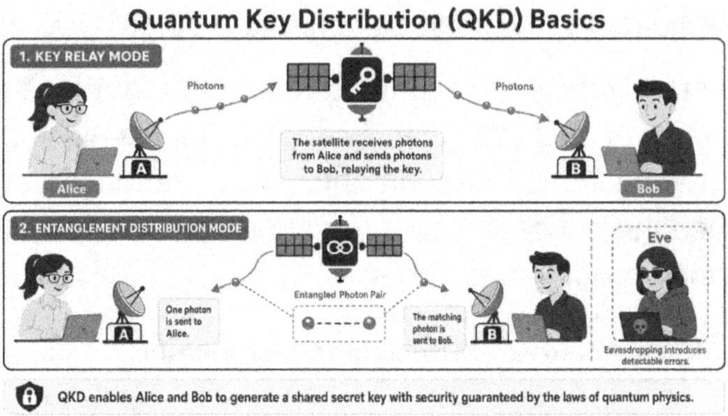

Diagram 8.3: Two satellite relay modes

In either scheme, the ground stations themselves can be far apart. The satellite overhead is the only thing connecting them during that pass. Because the satellite is in constant motion, crossing hundreds of kilometers of sky in just a few minutes, the key exchange must happen quickly and efficiently. Engineers must squeeze as many usable photons through the link as possible in each pass to accumulate a key long enough for practical encryption.

It is also possible to use a satellite purely as a classical relay for post-processing. The quantum exchanges occur separately between each ground station and the satellite. The classical error correction and privacy amplification steps, which you learned about in the BB84 chapter, are then completed with the satellite's assistance. The final output is a shared secret key at each ground station. From the user's perspective, the two stations have exchanged a quantum key over a very long distance, even though no single photon ever traveled directly from one station to the other.

Understanding this relay structure is important for the discussion of trust later in this chapter. But first, it helps to see what actual experiments have demonstrated, because the concept would remain theoretical without real-world proof that it works.

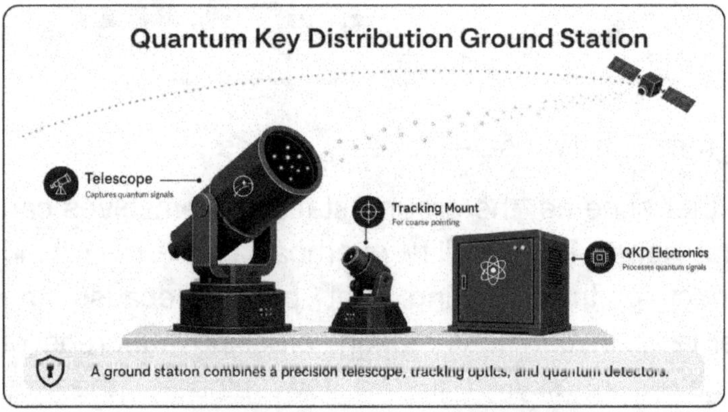

Diagram 8.4: Ground station architecture

The Landmark Demonstrations

For many years, satellite QKD was a promising idea that nobody had tested at scale. The physics was plausible, the calculations were encouraging, but until a working satellite actually demonstrated the exchange, it remained speculative. That changed when a Chinese space mission placed a specially equipped satellite into low Earth orbit and began a series of experiments that the scientific community watched closely.

The satellite carried onboard sources for generating quantum states and precision optics for directing those states toward ground stations far below. During its early mission, the satellite demonstrated QKD links to individual ground stations. The distances involved were far beyond what any fiber system had achieved. The satellite successfully delivered quantum keys to ground stations despite atmospheric turbulence, spacecraft motion, and the difficulty of holding a beam of single photons on a target the size of a dinner plate, seen from hundreds of kilometers above.

The most striking demonstration linked two ground stations that were separated by over a thousand kilometers, with the satellite serving as the relay. Neither station could have reached the other directly through fiber while maintaining useful key rates. Through the satellite relay, they exchanged a key, which was then used to encrypt and transmit an actual image over a classical channel. The quantum key had been used to secure a real communication, not just recorded in a laboratory notebook as a proof of concept. That step from demonstration to real use mattered enormously to the field.

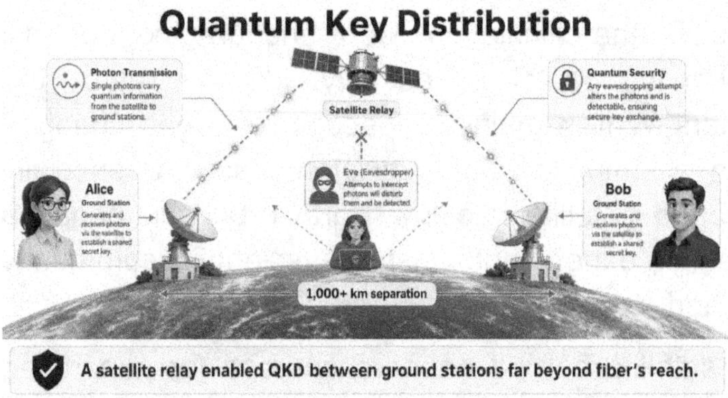

Diagram 8.5: The landmark two-station relay experiment

What made these demonstrations scientifically valuable was not just the distance but the rigor of the measurements. The teams involved published their results in peer-reviewed scientific journals, allowing the broader physics community to scrutinize the methods and conclusions. Independent researchers confirmed that the experiments were genuinely demonstrating quantum key distribution, not classical key exchange dressed up with quantum terminology. The no-cloning protection was real, and the keys were genuinely quantum in origin.

Later experiments from the same program extended the work further. Entanglement was distributed from the satellite to two ground stations simultaneously, with the entangled photon pairs surviving the trip through hundreds of kilometers of space and atmosphere. This demonstration opened the door to the entanglement-based relay approach described in the previous section, in which the satellite need not hold the key itself. The fact that entanglement could survive the journey from orbit to the ground was itself a significant result, as there had been genuine scientific debate about whether the fragile correlations between entangled particles could withstand it.

Other programs around the world took note. European space agencies began planning their own missions. Research groups in Canada, Japan, Singapore, and elsewhere proposed experiments of their own, ranging from small, affordable satellite buses carrying stripped-down QKD payloads to ambitious multi-satellite constellations. The Chinese demonstrations did not hand these programs a finished product, but they handed them something equally valuable: proof that the physics works and a benchmark to aim for.

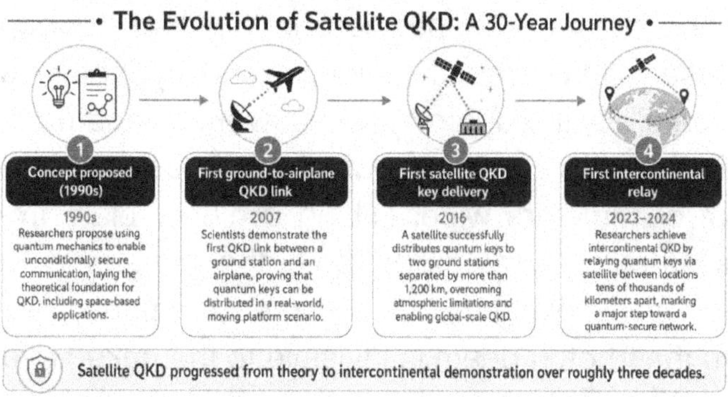

Diagram 8.6: Timeline of satellite QKD milestones

The Challenges of Pointing and Timing

Demonstrating that the physics works does not mean the engineering is easy. Anyone who has tried to photograph a fast-moving aircraft from the ground will have a small taste of the challenge involved. A satellite in low Earth orbit travels at roughly 7 kilometers per second. Seen from the ground, it crosses the entire sky in a few minutes. Holding a narrow beam of single photons on a telescope aperture that subtends a tiny angle of sky, all while both the satellite and the Earth-based telescope are moving, is one of the hardest precision engineering problems in the field.

The core difficulty is called pointing accuracy. A QKD satellite sends photons in a carefully focused beam, but even a focused beam spreads with distance. By the time the beam reaches the ground station hundreds of kilometers below, it has expanded to a footprint perhaps tens of meters across. The ground station telescope must be somewhere within that footprint. Conversely, if the ground station is sending photons up to the satellite, the beam from below must land on the satellite's receiver aperture, which might be only a few tens of centimeters wide. Hitting a dinner plate from 500 kilometers away while both the plate and the thrower are moving is an apt mental image, even though the photon beam is invisible to the naked eye.

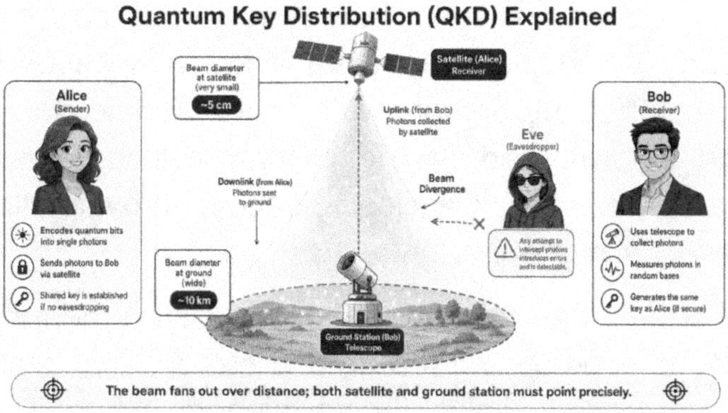

Diagram 8.7: Pointing and beam divergence

Engineers solve the pointing problem with a two-stage system. A coarse tracking mount keeps the telescope aimed roughly at the right part of the sky, using orbital predictions to determine where the satellite will be. A fine tracking system then takes over, using a bright reference laser beacon to lock on to the exact position in real time. The satellite sends down a strong classical laser pulse alongside the quantum signal. The ground station's tracking optics detect the bright beacon and use it to adjust the fine-pointing mirror many times per second, keeping the faint quantum beam centered on the detector. The quantum signal itself is too faint to track; it needs the beacon as a guide.

Timing is equally critical. Because the satellite is moving rapidly, the delay between sending and receiving a photon is constantly changing. Each photon must be assigned a precise timestamp so that the ground station and the satellite can match up which photons arrived and which were lost in transit. The timestamps must be synchronized to within a small fraction of a nanosecond, because at the data rates involved, a timing error of even a few nanoseconds would confuse the photons with noise. Atomic clocks on the satellite and at the ground station, cross-checked with the classical beacon, provide the necessary synchronization.

There is also the problem of Doppler shift. When the satellite is approaching the ground station, its signals are compressed slightly in wavelength; when it is receding, they are stretched. For classical communications, this is a minor nuisance easily corrected. For quantum signals in which the photon's polarization or phase carries the key bit, Doppler effects must be carefully tracked and compensated, or they will degrade the quantum bit error rate. The compensation is well understood, but it adds another layer of complexity to the ground station electronics.

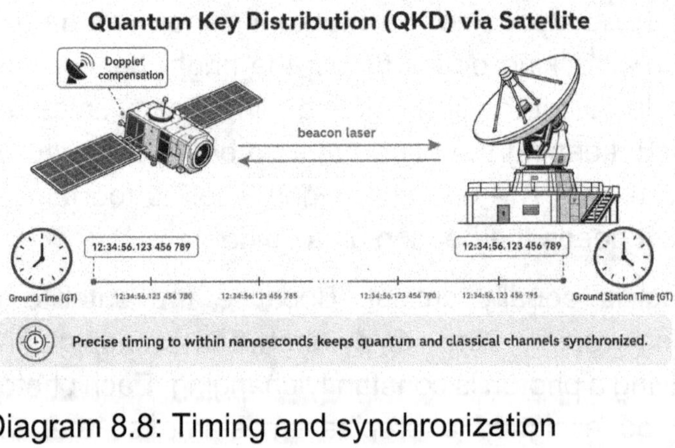

Diagram 8.8: Timing and synchronization

Despite these challenges, the engineering teams behind the landmark demonstrations solved them with enough reliability to achieve useful key rates. The pointing systems they developed are now informing a new generation of smaller, lighter satellites that aim to bring the cost of a QKD payload down from research-grade budgets to more in line with the economics of a conventional communications satellite. The hard part is mostly known; what remains is to build it faster and cheaper.

Weather, Daylight, and Brief Windows

Honest treatment of satellite QKD requires confronting its practical constraints head-on. The technology works, but not whenever you want it. Three factors limit when keys can actually be exchanged: the atmosphere's clarity, the sky's brightness, and the brevity of each satellite pass.

The clearest constraint is weather. Clouds block photons completely. A thick cloud deck between the satellite and the ground station means zero photons arrive at the detector, regardless of how well the satellite is pointed or how sensitive the detector is. This is not a solvable engineering problem in the usual sense; it is a meteorological one. A location that is cloudy 60 percent of the time loses 60 percent of its potential key exchange windows. Ground stations in regions with frequent cloud cover are less useful than stations in high deserts where the sky is clear most nights. In practice, the sites chosen for ground stations are often the same high-altitude observatories used by astronomers for the same reason.

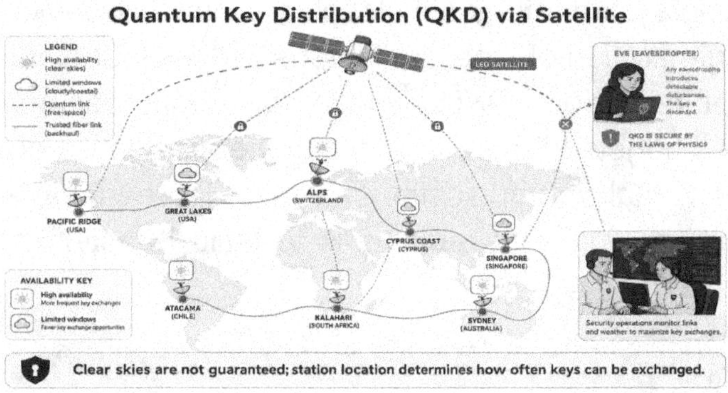

Diagram 8.9: Weather constraints on ground stations

Daylight is a related but distinct problem. The single photons used in QKD are faint by design. Daytime sky is bright by comparison, flooding the detector with background light that drowns out the signal. Early satellite QKD experiments operated only at night for this reason. Researchers have since developed narrow wavelength filters and time-gating techniques that allow some daytime operation, but daytime links still yield fewer usable photons per second than nighttime links. The practical reality is that most current satellite QKD happens after dark. Since a low-Earth-orbit satellite in a daylit orbit passes over the ground station during local daytime much of the time, this constraint further reduces the available windows.

The third constraint is the duration of each satellite pass. A satellite in low Earth orbit completes one orbit in roughly 90 minutes. From any given ground station, the satellite is above the horizon for only about 5 to 10 minutes per pass. Many passes occur during daylight or through clouds, leaving only a fraction that are usable for QKD. In a full 24-hour day, a ground station might have two or three usable passes from a single satellite. Each pass might yield enough photon detections to produce a key of perhaps a few hundred thousand bits under good conditions, using a high-performance system. That is enough to encrypt a substantial document, but it is not a continuous service running around the clock.

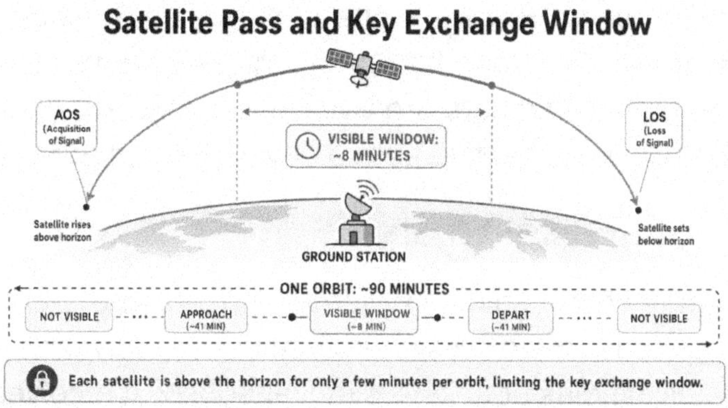

Diagram 8.10: The satellite pass window

The engineering response to these constraints is architectural. Instead of relying on a single satellite and hoping for good weather, designers plan for networks. Multiple satellites in different orbital planes ensure that at least one is overhead at any given station most hours of the day. Ground station diversity means that even if one site is cloudy, a nearby site in a different weather regime can serve the same user through a terrestrial fiber link connecting the two stations. The combination of a satellite constellation and an interconnected ground network is beginning to look like a resilient service rather than an opportunistic experiment.

There are also proposals to use higher orbits. A satellite in medium Earth orbit or geostationary orbit stays above the horizon much longer, giving much larger exchange windows. The tradeoff is distance: a geostationary satellite is about 36,000 kilometers up rather than 500 kilometers, which means the beam has much farther to travel and the photon loss is higher. Engineering a high-orbit QKD link is significantly harder, but researchers are working on it because the geometry is so attractive from a coverage standpoint.

Is the Satellite Trusted?

You may have noticed something in the relay description from the earlier section: in key relay mode, the satellite holds both keys simultaneously. It receives one key from ground station A and a separate key from ground station B. It then combines them to produce the shared secret. At the moment of combination, the satellite knows both keys. Any adversary who can compromise the satellite or capture its memory at that moment can recover the secret.

This is the same trusted-node problem you encountered in the previous chapter when reading about ground relay networks. The satellite, in key relay mode, is a trusted node in orbit. It is a particularly interesting trusted node because it is more difficult for an adversary to access than a ground-level building physically. You cannot just break into a satellite. But it is not impossible to compromise: sophisticated state-level adversaries might interfere with the satellite's software, intercept its classical communications, or tamper with it during manufacturing. Trust is not simply a matter of physical security.

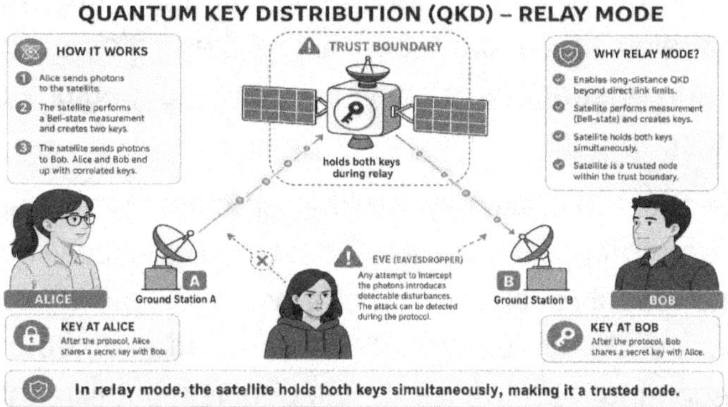

Diagram 8.11: The trusted satellite node problem

The entanglement-based approach, described earlier, removes this requirement. When a satellite distributes entangled photon pairs to two ground stations without storing or reading the correlated measurement outcomes, it does not know the key. The key exists only in the correlated measurement results held by the two ground stations. This is analogous to a postal service that delivers sealed, tamper-evident envelopes: the postmaster never opens them, so the postmaster need not be trusted with the contents. The entanglement-based satellite genuinely does not need to be trusted with the key, because it never holds it.

In practice, demonstrating fully untrusted entanglement-based satellite QKD at useful rates is harder than demonstrating trusted-relay satellite QKD.

The photon sources, detectors, and analysis methods required for entanglement-based QKD with full loophole-free verification are more demanding than those for the prepare-and-measure approach used in BB84-style relay systems. The landmark Chinese experiments demonstrated both modes, but the trusted-relay mode produced higher key rates in the early demonstrations simply because its equipment was more mature.

Researchers are actively working on closing this gap. New generations of entangled-photon sources that operate reliably in space and produce high-rate photon pairs are under development. Ground station detectors are becoming more efficient. The goal is a satellite that can distribute entangled pairs at a rate fast enough to produce useful keys without ever holding them. Some experimental programs have already demonstrated the entanglement leg of this system from orbit. Full end-to-end demonstrations at useful key rates, with the security of the entanglement properly verified, are the next milestone.

For now, if you are evaluating satellite QKD as a practical tool, you should ask which mode the system uses. A trusted-relay satellite offers real security advantages over classical key distribution, because an eavesdropper on the ground links will be detected. But the satellite operator must be fully trusted. An entanglement-based satellite removes that requirement, at the cost of lower current key rates and more complex verification. Knowing the difference lets you make an informed judgment about what guarantees you actually have.

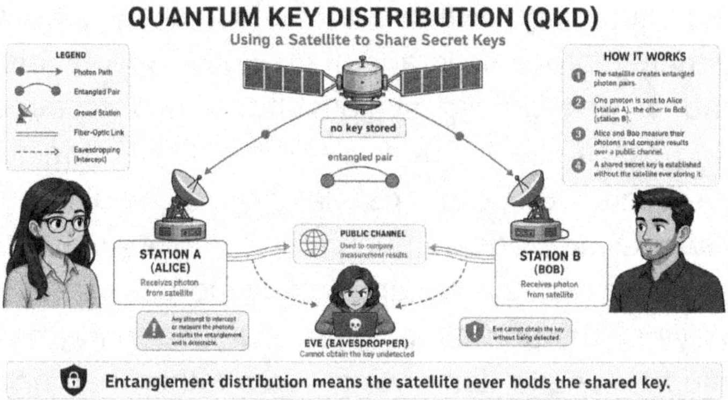

Diagram 8.12: Entanglement distribution removes satellite trust

The Coming Constellations

One satellite demonstrates a concept. A constellation turns a concept into a service. The transition from the former to the latter is what the next decade of satellite QKD is about.

A constellation is a coordinated network of satellites working together. In the context of communications, you may already be familiar with the concept from large low-orbit broadband networks that provide internet access in remote areas. The same orbital geometry that makes such networks useful for data also makes it useful for quantum keys: many satellites in different orbital planes ensure that from any point on Earth's surface, at least one satellite is above the horizon at almost any moment. The brief-pass problem largely disappears when you have enough satellites spread across enough orbital tracks.

For quantum key distribution, a constellation does more than improve coverage. It changes the key rate. Instead of one ground station getting two or three usable passes per day from a single satellite, it might receive dozens of passes per day from many different satellites. Over 24 hours, the cumulative key material can grow from a few hundred thousand bits to many megabytes, enough to provide real-time encrypted communications to demanding users rather than only in occasional batches.

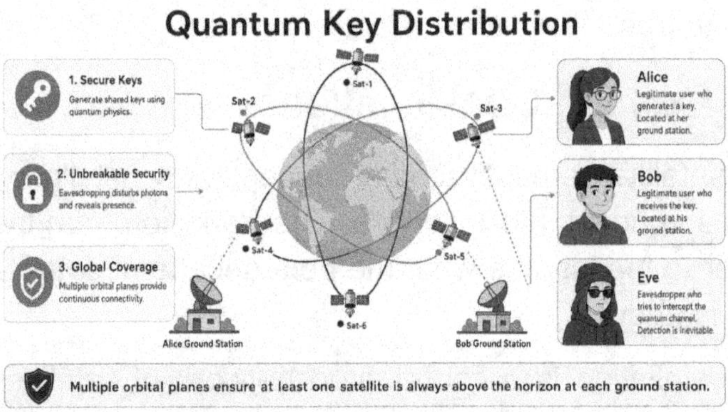

Diagram 8.13: A quantum satellite constellation

Several programs around the world are working toward this vision. European space and quantum-technology initiatives have published plans for a space-based component of a continent-spanning quantum network, with satellites serving as the long-distance bridges between national ground fiber networks. These plans call for a gradual build-up: early experimental satellites to prove the system designs, followed by operational satellites that provide actual service to government and commercial users. The timelines mentioned in public documents suggest initial service capability in the 2030s, with a more complete constellation following later.

The economics of this build-out matter. Early QKD satellites were research vehicles: expensive, custom-built, carrying one-of-a-kind instruments. The next generation aims to use standardized small satellites, sometimes called cubesats or microsats, which benefit from the same mass-manufacturing economics that have lowered the cost of commercial satellite communications. If a QKD payload can be packaged into a small satellite bus and launched as part of a rideshare mission, the cost per orbit drops dramatically. Several research teams have demonstrated that a stripped-down QKD transmitter can fit within the mass and power budget of a small satellite. Full results from these compact systems are beginning to appear in the scientific literature.

Security policy considerations are also driving the constellation programs. National governments in Europe, Asia, and elsewhere have identified sovereign quantum communications networks as strategic infrastructure, comparable to secure telecommunications cables or military communications satellites. The motivation is partly the harvest-now-decrypt-later threat discussed in earlier chapters: an adversary collecting today's encrypted government traffic could decrypt it years from now when quantum computers become powerful enough. Satellite QKD, as part of a broader quantum network, is seen as a hedge against that threat at the intercontinental scale where fiber QKD cannot reach.

Ground infrastructure must scale alongside the satellites. A constellation of dozens of satellites is only useful if the ground stations below them are equally capable. Automated ground stations that can track satellites without human operators on site, operate through mild atmospheric turbulence with adaptive optics, and connect into terrestrial QKD networks without manual intervention are needed. Prototypes of such automated stations have been demonstrated, and their designs are informing plans for the network's ground segments. The vision is a system in which a government agency, bank, or research institution in one country can request a quantum-secured key exchange with a counterpart in another country, and the network automatically schedules and executes the satellite passes.

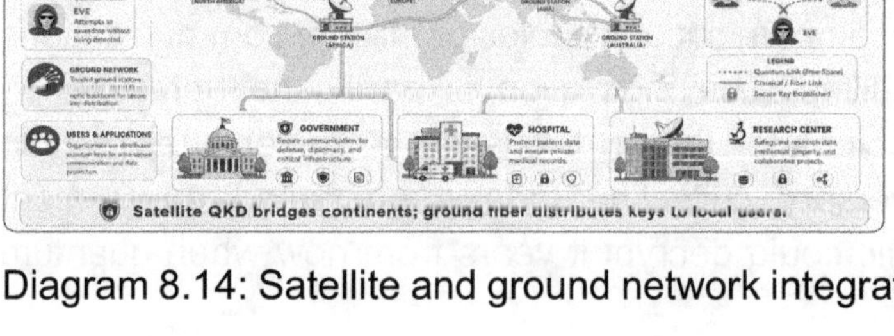

Diagram 8.14: Satellite and ground network integration

The constellation programs do not promise that satellite QKD will be simple or cheap in the near term. Each satellite has a limited operational life, typically five to ten years. Maintaining a constellation means a steady cadence of launches. The ground infrastructure investment is substantial. And the trusted-satellite problem, in most near-term deployments, will still rely on organizational trust in the satellite operators rather than purely on physics. These are honest assessments that the field itself acknowledges.

What the field does not doubt is that the direction of travel is clear. Fiber QKD has a distance ceiling that physics will not let us raise much further without quantum repeaters, which are still years away from practical deployment. Satellite QKD bypasses that ceiling today, with technology that already works. The demonstrations are real, the engineering is understood, and the programs are funded. From here, the question is not whether satellite QKD will scale up but how fast.

Key Concepts Checklist

• I understand why the near-vacuum of space and the thinness of the atmosphere make space a better photon medium than long fiber.

• I can describe how a satellite in key relay mode links two distant ground stations by sharing keys with each one and then combining them.

• I can describe how entanglement-based relay removes the requirement to trust the satellite with the key.

• I can name the landmark satellite QKD demonstration that showed intercontinental key exchange was possible.

• I understand why pointing accuracy and timing synchronization are demanding in satellite QKD systems.

• I know that weather, daylight, and short orbital passes limit when satellite QKD can operate.

• I understand that a satellite constellation improves coverage and cumulative key rates compared to a single satellite.

Questions to Explore

• Why does traveling vertically through 20 kilometers of atmosphere lose fewer photons than traveling horizontally through 200 kilometers of fiber?

• If a single satellite can only be above the horizon for 8 minutes per pass, what architectural choices would a network designer make to ensure reliable key delivery to users who need it every hour?

• A trusted satellite operator has significant power over the security of a relay-mode key exchange. Who do you think should operate such satellites, and what governance structures might provide oversight?

• What would need to be true technically, economically, and politically for satellite QKD to become as routine as commercial satellite broadband?

Try This / Reflect

Find a free satellite-tracking app or website (several widely used options are available for smartphones and browsers). Look up the passes of a low-Earth-orbit satellite visible from your location over the next few days. Set a timer when a pass begins and stop it when the satellite drops below the horizon. That brief window is all a quantum satellite has to deliver keys to your local ground station. Consider how few usable passes per day you might have, how clouds would eliminate some of those, and how a constellation of many satellites would change that picture. Then think about how ground fiber networks must serve as the backbone connecting different stations, so that keys accumulated at a clear-sky site can reach a nearby cloudy city.

Glossary Terms

• **Satellite QKD -** A method of distributing quantum keys using a spacecraft as an intermediate relay, bridging ground stations too far apart for direct fiber connection.

• **Ground station -** A site equipped with a precision telescope, tracking optics, and quantum photon detectors that exchanges quantum signals with a satellite passing overhead.

• **Free-space link** - A photon transmission path through open air or space rather than through fiber, used in both satellite QKD and shorter-range line-of-sight terrestrial QKD links.

• **Satellite pass** - The brief interval, typically a few minutes, during which a low-Earth-orbit satellite is above the horizon and within optical communication range of a ground station.

• **Constellation** - A coordinated network of multiple satellites in various orbital planes, designed to ensure that at least one satellite is above the horizon from any given location at almost any time.

• **Adaptive optics** - A technique that uses a rapidly adjusting mirror to correct the distortions that atmospheric turbulence introduces into an incoming photon beam, improving the signal received at the ground station.

Key Takeaways

• The near-vacuum of space absorbs almost no light, so photons survive a space crossing far better than a comparable journey through fiber.

• A satellite in low orbit can relay quantum keys between ground stations separated by thousands of kilometers, using either a trusted-relay or an entanglement-based approach.

• Landmark experiments have confirmed that satellite QKD works at intercontinental scale, with results published and independently verified.

• Weather, background light from the daytime sky, and short orbital pass windows all constrain when and how often keys can be exchanged, shaping how the technology must be deployed.

• Most current satellite QKD treats the spacecraft as a trusted node, but entanglement-based systems under active development can remove that trust requirement.

9 Who Is Using QKD Today

Overview

Quantum key distribution (QKD) sometimes gets portrayed as a laboratory curiosity that never quite makes it into the real world. That portrayal is wrong. Banks have installed QKD hardware between their trading floors and backup data centers to protect financial settlement data. Governments have deployed quantum-secure links to guard sensitive communications between ministries and military command posts. Power utilities, telecom carriers, and data-center operators have run pilots connecting critical sites with fiber that carries quantum keys alongside ordinary encrypted traffic. Healthcare institutions and research hospitals have begun to consider QKD for genomic archives whose secrecy must hold for a human lifetime. And several national governments, particularly in Asia and Europe, have built QKD into long-term infrastructure strategies, much as earlier generations built roads and power grids. This chapter surveys all of those real-world deployments and explains what they have in common.

Just as important as knowing who uses QKD is understanding why the list is still relatively short. The common thread running through every genuine adopter is a specific combination of three things: secrets that must stay confidential for many decades, communication routes that are fixed and important enough to justify the hardware investment, and budgets that can absorb the cost of specialized equipment and ongoing maintenance. Most organizations do not have all three at once. For the vast majority of everyday encryption needs, conventional methods and the newer class of post-quantum cryptography (PQC) algorithms are simpler, cheaper, and entirely adequate. Recognizing that gap is not a criticism of QKD. It is an honest description of a specialized, high-assurance tool that occupies a well-defined niche, one that will likely grow as quantum computers become more capable but that will never replace everyday encryption across the board.

Opening Scenario

Two buildings sit a few kilometers apart in a major financial district. The first is a bank's primary trading floor, humming with servers that process thousands of transactions per second. The second is the bank's backup data center, a quieter facility whose entire purpose is to hold an up-to-the-minute copy of everything happening in the first building. Between them runs a length of dedicated optical fiber. On that fiber, mixed in with the ordinary encrypted traffic, flows a continuous stream of single photons whose polarization states are being used to generate fresh cryptographic keys every few seconds. The bank is running a live QKD link.

The bank did not build this infrastructure because the technology is exciting, though it is. The bank built it because the data crossing that link carries settlement instructions, client positions, and interbank agreements that, if exposed even decades from now, could affect real people and real accounts. The executives who approved the budget understood something precise: the secrets on that link have a very long shelf life, the route is fixed and important enough to wire permanently, and the organization has both the technical staff and the financial resources to maintain specialized hardware. All three conditions were true, and so QKD made sense.

This chapter is about organizations that have made exactly that calculation. Some of them are banks. Some are ministries of defense. Some are national railway operators or electric-grid controllers. A few are hospital networks sitting on genomic databases. What they share is not a love of cutting-edge technology for its own sake. What they share is a specific security problem that QKD solves better than any alternative available today.

What This Chapter Will Cover

• Banking and finance: why financial institutions were among the earliest adopters and what their metropolitan QKD links look like in practice.

• Government and defense: how national governments use QKD to protect sensitive communications and why many treat quantum-secure infrastructure as a question of sovereignty.

• Critical infrastructure and data centers: the pilots and permanent links connecting power, telecom, and cloud-computing facilities, and why the fixed point-to-point architecture suits QKD so well.

• Healthcare and long-lived records: the connection between QKD's appeal and data that must stay confidential across a human lifetime, including genomic and health records.

• National and regional programs: the large coordinated initiatives in Asia and Europe that fold QKD into government infrastructure plans at a scale no single organization could achieve alone.

• The common thread among adopters: identifying the shared profile that predicts who is and is not a natural QKD candidate.

• Why most applications do not use it: an honest account of why QKD is overkill for most everyday encryption tasks and what alternatives fill that space.

Banking and Finance

Money is essentially information. A bank's most sensitive asset is not the physical cash in its vaults; it is the data describing who owns what, what trades were made at what prices, and what obligations one institution owes to another. That information has a long tail. A settlement record from today might be relevant in a legal dispute ten or twenty years from now. Client account details can affect people's lives for decades. Insider information about a major deal, if exposed after the fact, can still be used to construct profitable trading strategies or to embarrass parties to a negotiation.

Banks recognized this long-horizon problem earlier than most industries. When researchers began warning that a powerful quantum computer could eventually break the public-key encryption that protects most internet traffic, financial regulators in several countries quietly asked their largest institutions to start preparing. Some of those institutions looked at QKD as one option among several. A subset of them, those with short fixed routes between high-value sites and with the engineering resources to maintain specialized hardware, concluded that QKD was worth the investment.

The most natural fit is the metropolitan link: a dedicated fiber connection between two sites that are a few kilometers to perhaps a few tens of kilometers apart. In a large city with a dense financial district, it is common for a bank's primary data center, trading systems, and backup facilities to be close enough to connect with a single fiber run. That short distance is actually an advantage for QKD. Recall from earlier chapters that quantum signals weaken as they travel and cannot be amplified without disturbing the quantum states. A short, well-maintained fiber link loses very little signal and introduces very little noise, which means key generation rates are high and the system performs reliably. Banks with campuses spread across a single metropolitan area are therefore almost perfectly suited to QKD.

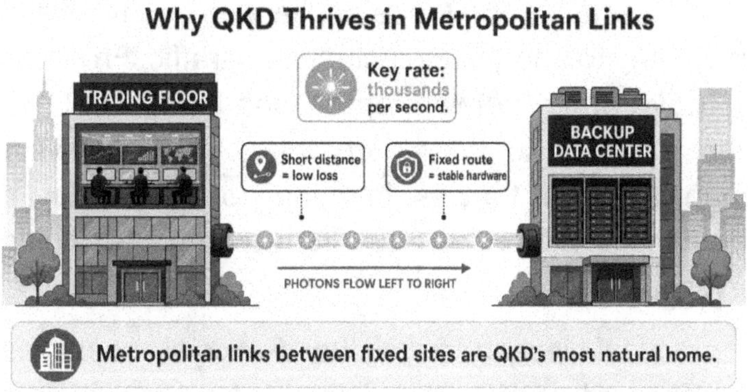

Diagram 9.1: A metropolitan bank QKD link

The financial sector has also been a proving ground because banks understand how to think about residual risk. A bank that processes a trillion dollars of transactions per year knows that even a tiny probability of a catastrophic data exposure represents an enormous expected loss. When risk managers run the numbers and compare them with the capital cost of a QKD installation, the math sometimes favors QKD, especially for links carrying the most sensitive data. This kind of actuarial reasoning, common in insurance and finance, is exactly the kind of thinking that drives technology adoption decisions in the sector.

Pilots in financial districts across Europe and Asia have demonstrated that QKD can be integrated into existing network management systems without disrupting normal operations. The quantum channel operates over a dedicated wavelength of light, separate from the classical channels that carry ordinary data. The QKD hardware generates a shared secret key, passes it to the bank's encryption appliances through a secure software interface, and those appliances use the fresh quantum key to encrypt the actual data traffic. From the perspective of the network operations team, the QKD system appears to be a very reliable key-delivery mechanism, one that is immune to the kinds of mathematical attacks a quantum computer would otherwise enable.

The practical lesson from the financial sector is straightforward. QKD adoption in banking is not random. It clusters around the specific conditions that make QKD work well: short fixed fiber routes, very high data sensitivity, long secrecy horizons, and organizational budgets large enough to sustain specialized infrastructure. Where those conditions exist, QKD is already a live technology, not a pilot, not a proof of concept, but a running system protecting real transactions today.

Government and Defense

Governments have secrets that last longer than any company's data. A defense intelligence assessment might remain classified for fifty years. Diplomatic cables describing negotiations between heads of state could remain sensitive for decades. Military command-and-control links carry information whose exposure could have strategic consequences that persist long after any individual transaction is forgotten. Governments have been considering long-lived secrets since before the age of computers, and it is no surprise that they are among the most active adopters of QKD.

The national security motivation for QKD is partly offensive and partly defensive. On the defensive side, a government that relies entirely on conventional public-key encryption faces a specific long-term threat: an adversary can collect encrypted government communications today and decrypt them later, once quantum computers become sufficiently powerful. This is the "harvest now, decrypt later" attack described in earlier chapters. For classified information that will remain sensitive for decades, a government cannot simply promise to upgrade its encryption when quantum computers arrive; by then, the stolen data will already be in an adversary's possession. QKD eliminates that threat on the links where it is deployed, because the keys are never transmitted in a form that can be harvested.

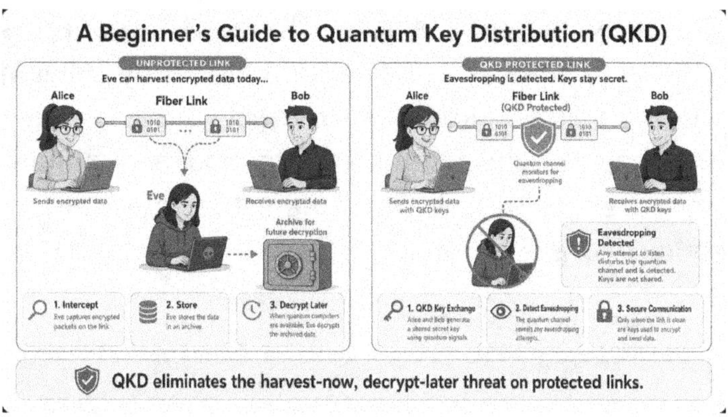

Diagram 9.2: Harvest now, decrypt later vs. QKD-protected link

On the offensive side, a government that builds quantum-secure communications infrastructure is also demonstrating technical capabilities with diplomatic and strategic value. Being able to say that sensitive communications are quantum-safe is not just a security posture; it is a signal to allies and adversaries alike about the sophistication of a country's technical infrastructure. Several countries have explicitly framed their QKD programs in terms of sovereignty: the ability to communicate securely without relying on encryption standards developed by foreign governments or on international bodies perceived as potentially compromised.

The installations that governments favor tend to be point-to-point links connecting fixed, known locations. Ministry headquarters to a backup secure facility. A parliament building to an executive office. A military command post to a regional operations center. These are all short, fixed routes connecting high-importance sites, exactly the architecture where QKD performs best. Some governments have gone further by building trusted-node networks that connect multiple sites within a capital city, effectively creating a quantum-secure intranet for the most sensitive government communications.

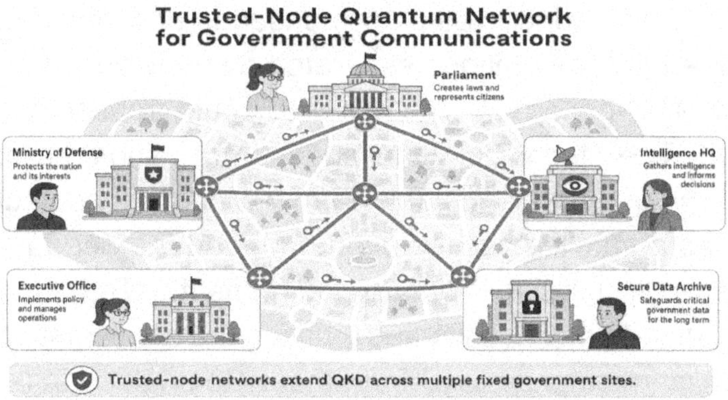

Diagram 9.3: A government trusted-node network

Defense establishments face a particular challenge that ordinary commercial organizations do not: the need for air-gapped or near-air-gapped communications that cannot be intercepted even in principle. A physical fiber carrying quantum signals between two secure facilities offers a very high degree of confidence about what is and is not happening on the link. A photon that is intercepted is disturbed, and the disturbance is detectable. That property resonates strongly with defense planners accustomed to treating adversarial interception attempts as a baseline assumption rather than an edge case.

Several governments have also invested in satellite-based QKD to extend quantum-secure communications over continental and intercontinental distances. A satellite in low Earth orbit, carrying QKD hardware and positioned to maintain line-of-sight contact with ground stations on opposite sides of a continent, can serve as a relay node for long-distance quantum key exchange. A major demonstration of this concept was conducted over thousands of kilometers between ground stations separated by a distance that no ground-based fiber QKD link could reasonably span. The technical challenges are substantial: atmospheric turbulence, the narrow window of contact time as a satellite passes overhead, and the need for extremely precise optical alignment. But the demonstrations have worked, shifting the conversation about long-distance QKD from theory to engineering.

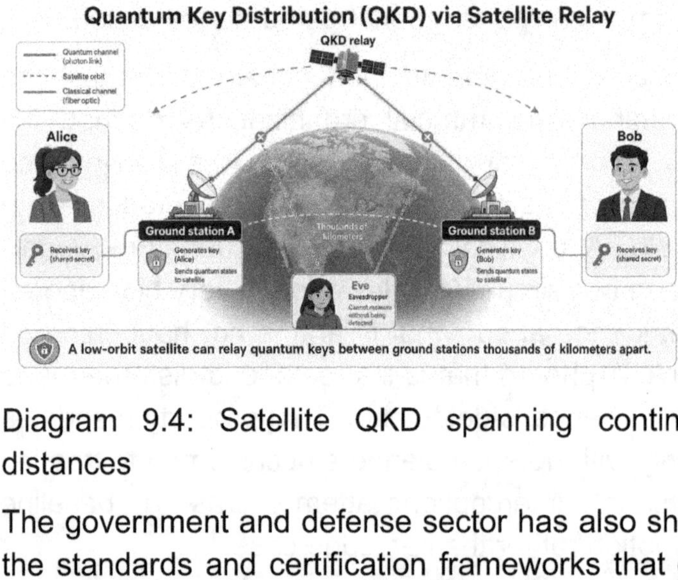

Diagram 9.4: Satellite QKD spanning continental distances

The government and defense sector has also shaped the standards and certification frameworks that other industries use when evaluating QKD. Several national security agencies have published technical guidelines describing the conditions under which QKD is acceptable, the minimum performance levels required for different classes of classified information, and the integration requirements for connecting QKD hardware to classified network infrastructure. These guidelines are not public in every country, but their existence reflects the seriousness with which defense establishments view the technology. When a government publishes a technical specification for how QKD must perform to protect military communications, it is sending a clear signal that this is a tool being prepared for operational use, not just academic study.

Critical Infrastructure and Data Centers

The phrase "critical infrastructure" covers a wide range of systems that modern societies depend on: electrical power grids, water treatment and distribution, telecommunications networks, financial clearing systems, transportation control systems, and the data centers that underpin cloud computing. All of these systems share an important characteristic. They are operated from fixed control facilities, connected to remote sites by fixed communication links, and dependent on the integrity and confidentiality of the instructions and data flowing across those links.

For a power grid operator, the communications links connecting a central dispatch center to substations and generation facilities are not merely a convenience. They are operational infrastructure. A compromised command signal could cause physical damage or outages affecting millions of people. The secrecy horizon for the operational data on those links is measured in decades because the infrastructure itself lasts that long. And the routes are, by definition, fixed: the dispatch center is where it is, the substations are where they are, and the fiber connects specific known locations.

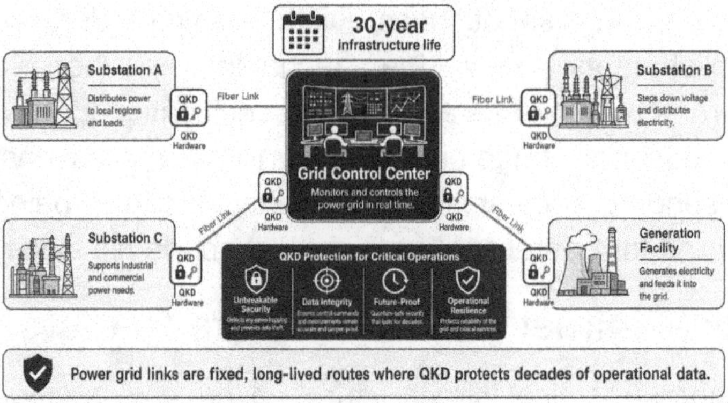

Diagram 9.5: QKD protecting a power grid control network

Telecommunications carriers represent another natural fit. A national telecom provider may operate a backbone network connecting dozens of cities, with critical management and control traffic flowing alongside customer data. The carrier's own operational data, including network maps, routing tables, and customer account databases, is valuable enough and long-lived enough to justify quantum-grade protection on the most sensitive links. Several carriers in Asia and Europe have run QKD pilots on portions of their backbone networks, using the quantum channel to protect the management plane while allowing customer traffic to continue on ordinary encrypted channels.

Data centers present a slightly different but related case. Large cloud computing providers and financial data center operators often operate multiple facilities within a metropolitan region, replicating data across them in real time for disaster recovery and load balancing. Those inter-data-center links carry everything: financial records, personal data, proprietary algorithms, medical records. They are also, as with banks, typically short and fixed. The distance between two data centers in the same metropolitan region is often short enough for QKD to operate at high key-generation rates and low error rates. Several data-center operators have piloted or deployed QKD on exactly these links, integrating quantum key delivery into the hardware encryption appliances that already protect inter-facility traffic.

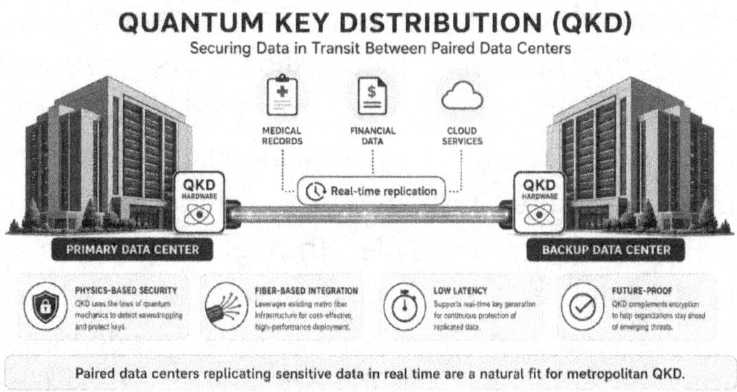

Diagram 9.6: QKD link between paired data centers

A simple observation about upgrade cycles reinforces the argument for QKD in critical infrastructure. A power substation might be built to last forty years. A data center might expect its physical infrastructure to last twenty years. The encryption protecting communications for those facilities needs to remain secure throughout the infrastructure's full lifetime, not just the next few years. If quantum computers capable of breaking conventional public-key encryption arrive in the 2030s or 2040s, infrastructure commissioned today needs to have accounted for that risk. QKD is one answer. Post-quantum cryptography is another. Some operators are choosing to deploy both as a belt-and-suspenders approach, using QKD on the most critical links and PQC across the broader network.

One of the genuinely interesting developments in critical infrastructure is the emergence of city-scale QKD testbeds. Several municipalities, particularly in Asia, have established metropolitan-area QKD networks that connect multiple government and critical-infrastructure sites within the city. These networks are not just pilots. They are live operational systems used by real organizations for real communications. The city itself serves as a natural boundary for a first generation of quantum-secure infrastructure, with inter-city or national extension possible through trusted-relay nodes or satellite links. These urban networks are proving grounds for the operational procedures, maintenance practices, and integration standards that will eventually allow QKD to scale from city pilots to national infrastructure.

Healthcare and Long-Lived Records

Health data is different from most other kinds of sensitive information in one important way: it is personal for life. A credit card number can be replaced in a few days. A password can be changed in minutes. But a person's medical history, genetic profile, and genomic sequence cannot be replaced. They describe that person permanently. If genomic data is exposed and then decrypted by a future quantum computer, the exposure cannot be reversed. The individual cannot change their genome any more than they can change their fingerprints.

This is the "shelf life of a secret" concept from an earlier chapter brought to its sharpest point. Most secrets lose their value over time, becoming less sensitive as circumstances change. Genomic data does the opposite. It remains just as sensitive fifty years from now as it is today, and arguably more so as genetic analysis tools improve and the amount of personal information extractable from a genome continues to grow. A hospital or research institution that stores genomic data today is holding a secret that could harm real people if exposed in 2060 or 2080. That institution bears responsibility for protecting it across that entire time horizon.

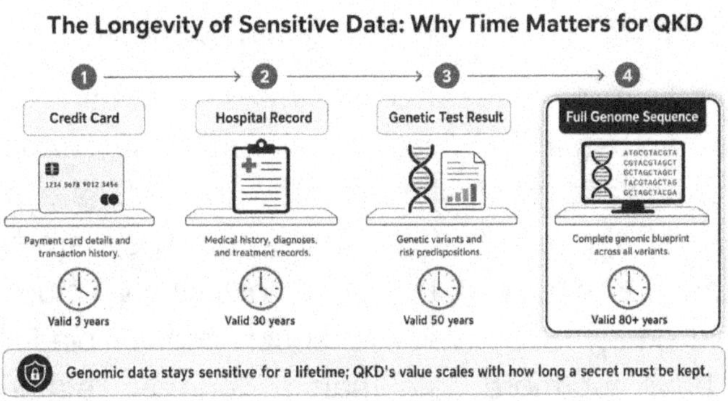

Diagram 9.7: The shelf life of health and genomic records

Hospitals and medical research institutions face a specific institutional challenge. They often must retain patient records for legally mandated minimum periods that extend for decades. In many jurisdictions, clinical trial data must be retained for 15 to 25 years. Genomic research data may need to be retained indefinitely to support follow-up studies. The institutions holding this data are therefore in a position to commit to secrecy that extends far beyond any foreseeable technology upgrade cycle.

QKD does not solve the entire problem of long-term health data security. A hospital's entire network cannot realistically be protected by QKD links in the way that a pair of data centers can. But for the specific links carrying the most sensitive genomic and clinical research data, particularly links between a research hospital and its secure archive facility, or between a genomic sequencing center and a national biobank, QKD is a technically sound option. This fixed-route, short-distance architecture suits QKD so well that it maps naturally onto such institutional connections.

The healthcare case also illustrates a pattern that appears across all the sectors covered in this chapter. The organizations most interested in QKD are not necessarily those with the most data. They are the organizations whose data is both extremely valuable and extremely long-lived. A social media platform might have far more data than a genomic research institute, but most of that data loses its sensitivity relatively quickly. A genomic archive may hold fewer records, but each record is permanently sensitive. That distinction is what drives QKD adoption, not the sheer volume of data but the combination of value and durability.

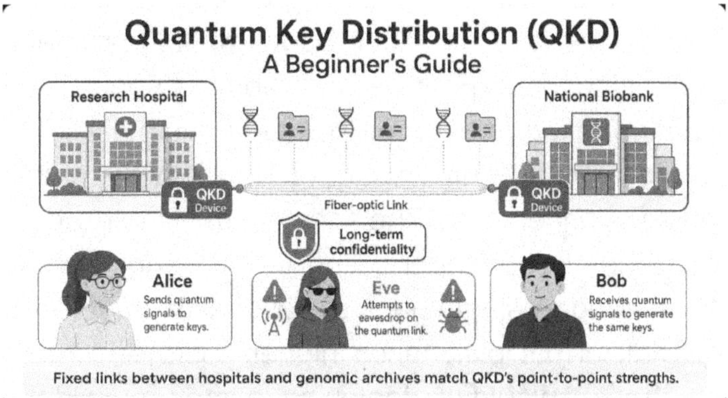

Diagram 9.8: Hospital-to-biobank QKD link

Healthcare institutions have been slower than banks and governments to deploy QKD, largely for practical reasons. Hospitals typically have smaller technology budgets than national banks or defense ministries, and they must balance security investments against clinical and research priorities. But as the cost of QKD hardware gradually decreases and regulatory pressure on long-term health data protection increases, the healthcare sector is an increasingly plausible market for QKD deployments. Several national health systems and large research consortia have begun feasibility studies, and the groundwork being laid by government and finance sector deployments in terms of integration standards, operational procedures, and maintenance training will make it easier for health sector organizations to follow.

National and Regional Programs

Individual organizations deciding to buy and install QKD hardware face a difficult bootstrapping problem. The technology is expensive and requires specialized engineering expertise to deploy and maintain. The market for QKD equipment is still relatively small, which keeps prices high. The ecosystem of integrators, managed service providers, and trained engineers is still developing. A single bank or hospital trying to build a QKD link on its own faces higher costs and more friction than it would if deploying a mature, widely-understood technology.

National programs change this picture significantly. When a government commits to building a national quantum-communication backbone and funds the infrastructure, training, and standardization work at a national level, it transforms QKD from something an individual organization must acquire and operate alone into something that organizations can access as a service or connect to as part of a larger network. Several countries have taken exactly this approach, with results that are changing the practical landscape of QKD deployment.

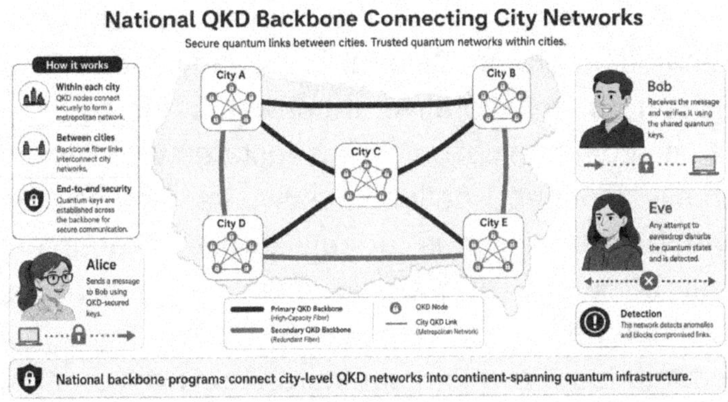

Diagram 9.9: A national QKD backbone with city nodes

The most extensive national QKD program to date has been built in Asia. A large country in that region has invested heavily in building a national quantum-communication backbone spanning thousands of kilometers, connecting dozens of cities and extending to major government and financial centers. The backbone uses trusted relay nodes (where keys are decrypted and re-encrypted at intermediate stations rather than remaining quantum-encrypted end-to-end) to extend the range beyond what any single fiber segment can support. The network has been used for real communications by government agencies, banks, and other organizations. It also served as the ground segment for a satellite QKD demonstration, extending the network's reach beyond the country's borders. This program demonstrated something important: that QKD is not confined to short metropolitan links but can, with sufficient investment in infrastructure, be scaled to national and even international scope.

European countries have also moved forward with coordinated programs. Several European nations have signed on to collaborative frameworks for building quantum communication infrastructure that would eventually connect national networks across the continent. The vision is a pan-European quantum internet backbone that would enable organizations across countries to exchange quantum keys through a shared, standardized infrastructure. The work is at an earlier stage than the Asian programs, but the political commitment and the funding mechanisms are in place. Several pilot corridors connecting major cities within individual countries have already been completed.

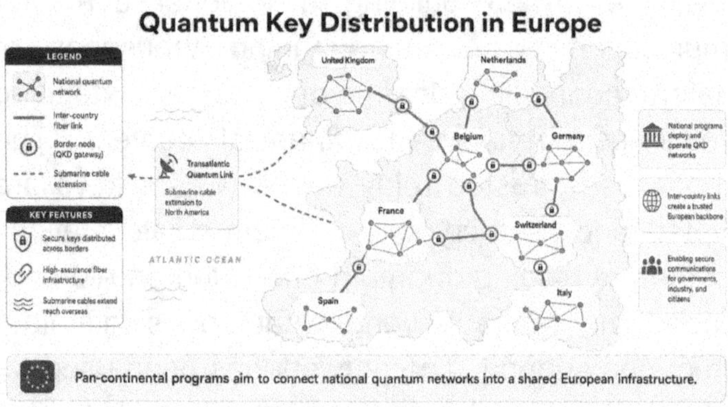

Diagram 9.10: A pan-European quantum backbone concept

National programs have also driven progress in standardization. When a single country decides to build a national network, it is forced to make decisions about interface specifications, key management protocols, security certification requirements, and operational procedures that individual organizations would otherwise each resolve differently. The standards that emerge from large national programs eventually become the foundation for commercial products and interoperability frameworks that allow smaller organizations to participate. Several national metrology institutes and government standards bodies have published technical specifications for QKD that are directly shaped by their countries' program experience.

The geopolitical dimension of national QKD programs deserves direct attention. Quantum communication infrastructure is infrastructure in the same sense that roads and power grids are infrastructure. Once built, it shapes the economic and strategic possibilities for the country that has it. A country with a national quantum backbone can offer organizations on that backbone a level of communication security unavailable elsewhere. That capability has economic value, because organizations dealing in highly sensitive information will prefer to operate where they can access it, and strategic value, because the government controlling the backbone knows which organizations are using it and can offer it as a service to allies. Several governments have been explicit that their QKD programs are not just security investments but industrial policy and strategic positioning.

The Common Thread Among Adopters

Looking across banking, government, critical infrastructure, healthcare, and national programs, a clear profile of the natural QKD adopter emerges. It is not defined by industry sector. It is defined by the specific combination of characteristics that make QKD's costs worth its benefits.

The first characteristic is secrets with a very long confidentiality horizon. Financial settlement records, government communications, power grid operational data, genomic sequences, and military intelligence assessments all share the property that their sensitivity persists for decades. The longer a secret needs to stay secret, the more valuable it is to protect it with an encryption scheme that resists attacks from future computers, not just today's.

The second characteristic is fixed, important routes. QKD works on dedicated point-to-point or multi-hop fiber links. It does not work on mobile devices, with occasional connections, or over dynamically routed internet traffic. Organizations that have specific, identifiable routes between fixed sites, routes that carry their most sensitive data, are the ones that can actually deploy QKD hardware productively. A bank with two permanent sites in the same city fits this description perfectly. A company whose employees connect from home on laptops does not.

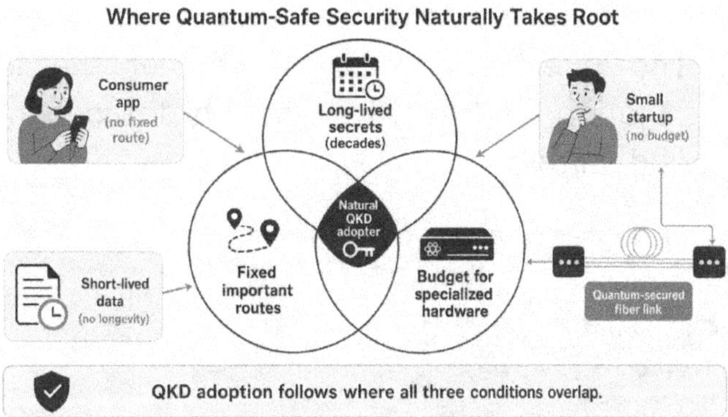

Diagram 9.11: The adopter profile: three overlapping conditions

The third characteristic is the budget and organizational capacity to sustain specialized hardware. A QKD installation is not a commodity product you can order, plug in, and forget. It requires trained engineers to maintain optical alignment, replace components, monitor key generation rates, and integrate the quantum layer with the classical encryption and key management infrastructure. Organizations that already operate specialized technical infrastructure- banks with network operations centers, defense establishments with their own engineering corps, national carriers with fiber maintenance teams- have the institutional capacity to take on QKD as an additional technical responsibility.

When all three characteristics are present together, QKD is a rational choice. When one or more is missing, the case for QKD weakens sharply. This is not a criticism of QKD; it is a description of a specialized tool. A laser cutter is excellent for cutting metal with precision, but you would not buy one to cut paper at home. The question is always whether the tool matches the job, and for QKD, the job description is specific and narrow.

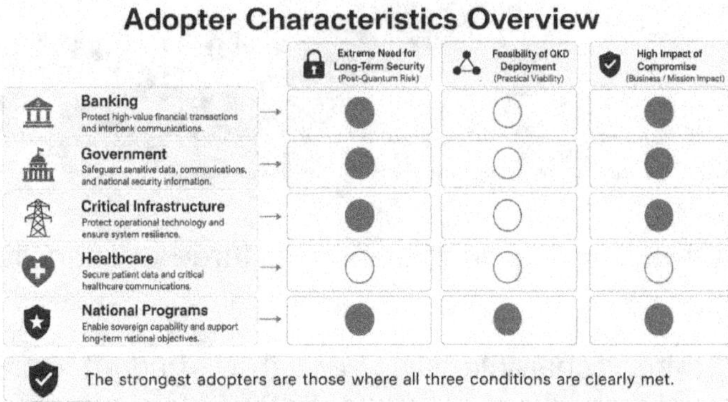

Diagram 9.12: Mapping sectors to the adopter profile

Understanding the adopter profile also helps you think about where QKD will grow next. As quantum computers become more capable and the threat to conventional encryption becomes more concrete and near-term, the set of organizations for which the first characteristic, long-lived secrets at risk from quantum decryption, is salient will expand. As QKD hardware improves and costs come down, the third characteristic, budget and capacity, will become easier to satisfy for a wider range of organizations. The second characteristic, fixed important routes, is structural and will not change for most organizations, but it does not need to. QKD will expand by going deeper into the sectors where it already fits, and by extending its geographic reach through national and international backbone programs, rather than by becoming a universal replacement for conventional encryption.

Why Most Applications Do Not Use It

It would be easy to walk away from this chapter thinking that QKD should be the default choice for any organization that cares about security. That impression would be wrong, and being honest about it is important. The vast majority of digital communication today does not require QKD and will not require it even when quantum computers arrive.

Consider the range of things you encrypt every day: your connection to a bank's website, a video call with a colleague, a file you upload to cloud storage, a text message to a friend. All of these involve encrypted communication. But think about what those secrets actually are. Your banking session uses a key that lasts a few minutes and protects a transaction that will be reflected in your account balance within seconds. The sensitive data has a very short useful lifetime. The idea that an adversary would record your encrypted banking session today, store it for twenty years, and then decrypt it with a quantum computer to find out what you looked at in your checking account on a Tuesday morning is not a realistic threat model. Even if the attacker succeeded, the information would be useless.

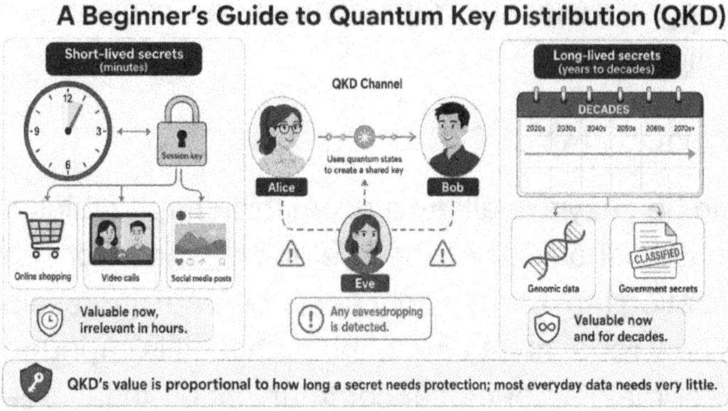

Diagram 9.13: Short-lived vs. long-lived secrets: the QKD relevance gap

The other major reason most applications do not use QKD is practical rather than conceptual. QKD requires dedicated fiber. The internet does not work that way. Internet traffic is dynamically routed through shared infrastructure, passing through dozens of intermediate routers and switches on the way from sender to receiver. There is no dedicated fiber between your laptop and a server in another country. Even if QKD hardware could be miniaturized and made affordable for consumer use, the fundamental architecture of the internet would make it impossible to deploy QKD on general internet traffic without a complete redesign of how global networks are built. That redesign is not happening, and there is no reason to pursue it to satisfy a security requirement that conventional and post-quantum methods already meet.

Post-quantum cryptography deserves special mention here because it is the most direct alternative to QKD for the general market. PQC algorithms are mathematical schemes designed to be hard to break even for a quantum computer. They run on existing hardware, require no dedicated fiber, work across the internet, and can protect mobile devices, IoT sensors, embedded systems, and cloud services. They have been through years of public cryptographic scrutiny, and national standards bodies have standardized several of them. For the overwhelming majority of encryption tasks, PQC provides quantum resistance without the operational complexity and hardware constraints of QKD. The honest summary is that PQC and QKD are complementary tools addressing different parts of the problem. PQC protects the broad, distributed, mobile, and commodity communications landscape. QKD protects the narrow set of highest-assurance, fixed-route, long-horizon use cases where its physical security guarantee matters most.

Even among organizations that have already deployed QKD, the technology typically protects only a small fraction of their total communications. A bank with a QKD link between its two main data centers still uses conventional TLS encryption for its customer-facing website, its employee email, its internal collaboration tools, and most of its network management traffic. The QKD link is a specialized, high-security layer for the most critical data flows, operating in parallel with a much larger infrastructure of conventional and post-quantum security measures. This tiered approach, using the right tool for each level of sensitivity, is exactly how mature security programs manage costs and complexity. It is not a compromise. It is good engineering.

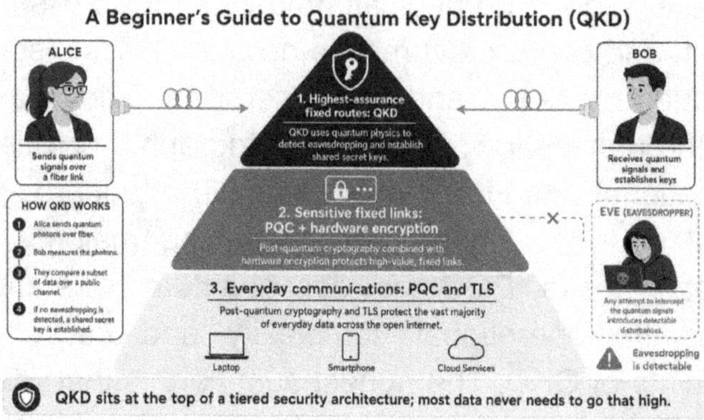

Diagram 9.14: Tiered encryption: QKD for the top, PQC for the rest

There is one more reason why most applications do not use QKD, and it is worth stating plainly. For many organizations, the threat of quantum computers breaking their encryption is not yet urgent enough to justify the investment. Post-quantum algorithms have been standardized and are being rolled out gradually across internet infrastructure. For an organization whose most sensitive data is not particularly long-lived, the combination of conventional encryption today and a planned migration to post-quantum algorithms in the next few years is a reasonable and adequate security strategy. QKD represents a more demanding alternative that makes sense only when the cost of any encryption failure, even a future one, is unacceptably high. Most organizations are not in that position. That is not a failure of QKD; it is a sign that QKD occupies exactly the niche it should occupy.

Key Concepts Checklist

• I can explain why banks were early QKD adopters and what specific conditions made metropolitan QKD links a natural fit for financial institutions.

• I understand why governments treat quantum-secure communication as a strategic and sovereignty issue, not just a technical security feature.

• I can describe a critical infrastructure use case for QKD and explain why fixed point-to-point links suit the technology's architecture.

• I can connect QKD adoption to the shelf life of a secret and explain why genomic and long-lived health data raises the case for quantum-secure links.

• I understand that large national programs in Asia and Europe are driving QKD adoption at a scale individual organizations could not achieve alone.

• I can identify the three conditions that define the natural QKD adopter: long-lived secrets, fixed important routes, and budget for specialized hardware.

• I can explain honestly why most everyday encryption tasks do not need QKD and what alternatives, particularly post-quantum cryptography, address those needs instead.

Questions to Explore

• What do all the natural QKD adopters have in common, and does your own organization share any parts of that profile? If you found a match, which of the three conditions would be hardest to satisfy?

- Why might a government view quantum-secure communication as a sovereignty issue rather than just a security feature, and how does that framing change the kind of investment it is willing to make?
- If QKD is overkill for most uses, why does it still matter that some organizations deploy it? What does a working QKD deployment accomplish for the broader ecosystem beyond protecting its own data?
- How would you decide whether a given secret justifies the cost of QKD? What questions would you ask, and what answers would push you toward QKD versus post-quantum cryptography?

Try This / Reflect

Search for a recent news story about a real QKD deployment. Look for one involving a bank, a government agency, or a city-scale pilot network. As you read, try to identify three things: who the users are, what specific category of data they are protecting, and what reason the story gives for choosing QKD over ordinary encryption. Then ask whether the same reasoning would apply to the most sensitive data your organization holds; if it would not, try to pin down exactly which of the three adopter conditions is missing.

If you want to go a step further, sketch a simple map of the two or three most sensitive data flows in an organization you know well. Mark where the data starts, where it ends, and how far apart those points are. Then ask whether those flows are fixed point-to-point connections or dynamic routed connections. That single question will tell you a great deal about whether QKD could ever be practical for that organization.

Glossary Terms

• **Use case -** A specific real-world application of a technology, defined by who uses it, what problem it solves, and why that solution is appropriate for that context.

• **Metropolitan link -** A short, fixed QKD fiber connection between two sites within a city or metropolitan area, typically the most practical and cost-effective deployment scenario for QKD.

• **Critical infrastructure -** The essential systems that modern societies depend on for basic functioning, including power grids, water systems, telecommunications networks, financial clearing systems, and transportation control systems.

• **High-assurance -** A level of security strong enough for the most sensitive and long-lived data, where the cost of any failure is unacceptably high and conventional security measures are not considered sufficient.

• **National program -** A government-led initiative to plan, fund, and build quantum communication infrastructure at a national scale, transforming QKD from a technology individual organizations must adopt alone into shared infrastructure others can connect to.

• **Post-quantum cryptography (PQC) -** A class of mathematical encryption algorithms designed to resist attacks from quantum computers, running on ordinary hardware without dedicated fiber, and covering the broad range of communications that QKD cannot practically reach.

Key Takeaways

• QKD is genuinely deployed today in banking, government, critical infrastructure, and selected healthcare and research settings, making it a real operational technology rather than a laboratory demonstration.

• The natural QKD adopter has three characteristics in combination: secrets that must stay confidential for decades, fixed important routes that can be physically wired, and the organizational budget and capacity to maintain specialized hardware.

• Much of the most significant QKD adoption is driven by national and regional government programs, which solve the bootstrapping problem by building shared infrastructure that individual organizations can connect to.

• QKD's appeal scales directly with how long a secret needs to stay secret; the longer the required confidentiality horizon, the stronger the case for quantum-physical security guarantees.

• For the vast majority of everyday encryption needs, QKD is overkill; post-quantum cryptography provides quantum resistance across the broad landscape of distributed, mobile, and commodity communications where QKD's fixed-link architecture cannot reach.

10 The Honest Limitations

Overview

Quantum key distribution (QKD) is one of the most elegant ideas in modern security. The physics is real, the guarantees are meaningful, and the technology has moved from laboratories into fiber networks spanning hundreds of kilometers. But a book that spends nine chapters explaining QKD's promise owes you, in this tenth chapter, an equally careful account of what QKD cannot do, where it has been defeated in practice, and why serious institutions remain cautious about deploying it. This chapter is that account. It does not aim to shake your confidence in QKD as a concept; it aims to give you the specific, honest map of the terrain so that you can read any future claim, whether glowing or dismissive, with a clear head.

The limitations covered here fall into a handful of distinct categories. First, QKD cannot, by itself, tell you who is on the other end of the quantum channel; it secures the link but does not identify the person. Second, it requires dedicated physical hardware and usually dedicated fiber or line-of-sight paths, so it cannot be packaged as software or delivered as a cloud service. Third, its long-distance form still depends on trusted relay nodes that see the raw key, re-exposing the very secret QKD was meant to protect. Fourth, real QKD hardware has been attacked and defeated in laboratory experiments, not by breaking the physics, but by exploiting imperfections in the devices that implement it. Fifth, QKD is expensive, has range limitations, and can be silenced by an attacker who disrupts the quantum channel. Sixth, major national security agencies in multiple countries have raised formal objections to QKD for sensitive deployments, and their reasoning deserves a fair hearing. And seventh, the nations and researchers actively building QKD networks have thoughtful answers to those objections. By the end of this chapter, you will be able to hold both views simultaneously and decide for yourself.

Opening Scenario

Imagine you pick up two articles about QKD on the same afternoon. The first is a press release from a technology company. It opens with the words "unhackable quantum encryption" and promises that its new product "makes data breaches a thing of the past." The graphics show padlocks with glowing quantum auras. The second article is a dry advisory from a government security body. It recommends against deploying QKD for most high-stakes uses, citing concerns about authentication gaps, hardware vulnerabilities, operational complexity, and cost. It prefers a different approach entirely, one based on mathematics rather than physics.

You feel confused. Both sources seem serious. One claims a breakthrough; the other recommends caution. They cannot both be entirely right, and yet both probably contain something true. The press release is describing the physics, which genuinely is remarkable.

The advisory describes the engineering and operational reality, which is genuinely complicated. These are two different conversations about the same technology, and the gap between them is where misunderstanding lives.

This chapter is built to close that gap. We will take the limitations seriously, one by one, in plain language, without turning any of them into a reason to dismiss QKD entirely or, on the other side, into a minor footnote. The goal is precision. A limitation that you understand clearly is not a reason for despair; it is a constraint you can work with. An exaggerated promise you believe is a trap waiting to spring.

What This Chapter Will Cover

• Why QKD does not solve authentication, and what that gap means in practice

• Why QKD requires special hardware and cannot be delivered as software or a cloud service

• The trusted-node problem revisited as a first-class limitation, not a side note

• Implementation attacks on hardware have defeated how real QKD systems work

• The practical limits of cost, distance, and denial of service

• What the skeptics argue, and why their reasoning is worth taking seriously

• What the adopters argue back, and how QKD and post-quantum cryptography can work together

It Doesn't Solve Authentication

Let us start with the limitation that surprises people most, because it sounds so counterintuitive. QKD creates a secret key between two parties in a way that is protected by the laws of physics. If anyone tries to observe the quantum signals used to build that key, they disturb those signals, and the disturbance shows up in the data. Alice and Bob detect the eavesdropping and abort the session. This is the core guarantee, and it is real.

But here is the question the guarantee does not answer: How do Alice and Bob know they are actually talking to each other?

Think about it this way. You call your bank. The conversation that follows is encrypted. But before you get to the encrypted part, something happens: the bank's website presents a digital certificate that your browser checks, or the phone system routes your call to a number you trust, or a voice robot reads you a code that you type into an app. All of these steps are authentication. They are the part of the system that answers the question "Is this really my bank, and am I really your customer?" Encryption and authentication are two distinct tasks, and both are necessary.

QKD handles encryption brilliantly. It generates a key so fresh and so physics-protected that even a future quantum computer cannot crack it retroactively. But

QKD does not, on its own, perform authentication. It builds a secure tunnel, but it does not verify who is at either end.

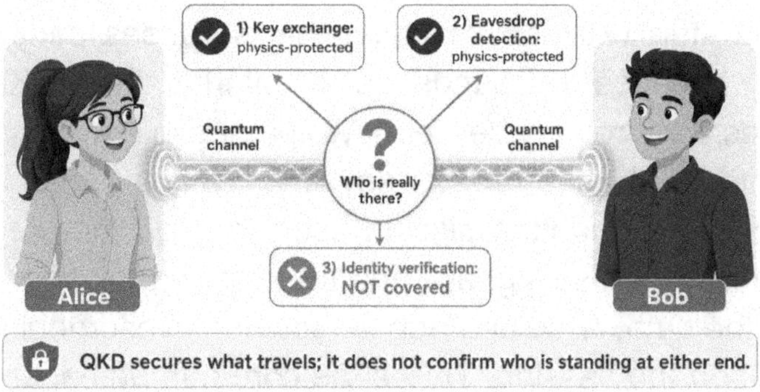

Diagram 10.1: The authentication gap in a QKD session

Why does this matter so much? Consider the man-in-the-middle attack. In this attack, a clever adversary, call her Eve, inserts herself into the middle of the setup phase before the quantum channel is used. She intercepts Alice's initial messages and impersonates Alice to Bob, while simultaneously impersonating Bob to Alice. Now Eve runs two separate QKD sessions: one with Alice and one with Bob. Each session is perfectly secure by the laws of physics. Alice believes her key is shared with Bob. Bob believes the same. But in reality, both keys belong to Eve. Eve can now decrypt Alice's messages, read them, re-encrypt them for Bob, and pass them along. Neither Alice nor Bob detects anything wrong, because their individual quantum channels are undisturbed.

This is not a theoretical ghost story. It is a real vulnerability and the primary reason QKD must always be combined with some form of pre-shared authentication. In practice, Alice and Bob need a small amount of shared secret before the QKD session begins. They use that initial secret to authenticate the classical control messages that accompany the quantum exchange. If no such pre-shared secret exists, the man-in-the-middle attack is possible.

But notice the subtle trap in that solution. If Alice and Bob need a pre-shared secret to authenticate QKD, then they already needed to solve the key distribution problem once, by some other means, to get QKD running. This is not a fatal flaw; it is a well-understood bootstrap problem. The pre-shared authentication secret does not need to be long, frequently refreshed, or quantum-protected. But it does need to exist. QKD is not a self-contained answer to the question "How do two strangers share a secret safely?" It is a powerful tool for refreshing and extending a secure relationship that was already established by some other means.

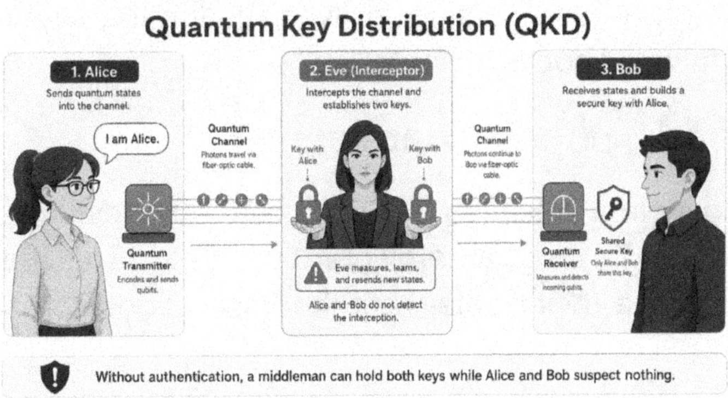

Diagram 10.2: The man-in-the-middle attack on unauthenticated QKD

There is a second, more subtle point about authentication that is worth understanding. Many post-quantum cryptography (PQC) algorithms, the math-based alternatives to QKD that we will discuss later in this chapter, are also public-key systems. That means they allow two strangers to authenticate each other without any prior shared secret. This is a genuine operational advantage that QKD currently lacks. A QKD system deployed from scratch between two organizations that have never met requires an initial secret-sharing ceremony of some kind: a trusted courier, a pre-shared password, or some other mechanism. For large networks where new nodes are constantly being added, this can become a significant operational burden.

None of this means QKD is broken or useless. It means QKD is a specialized tool that solves one specific problem extremely well, and that it depends on adjacent tools to solve the problems it does not touch. A lock on a door is valuable even though it does not confirm who knocks.

It Needs Special Hardware

Turn now to a limitation that is easy to underestimate: QKD is, at its core, a hardware technology. This may sound like a minor detail, but the implications spread far.

Classical cryptography, the kind that protects your browser traffic right now, can be implemented entirely in software. You download an app, or your browser gets an update, and a new encryption algorithm is running on your laptop without any change to your physical hardware. This is enormously powerful from a deployment perspective. When a classical algorithm is found to be weak, the cryptographic community can push out software updates to billions of devices within days or weeks. The infrastructure does not change; only the code does.

QKD cannot work this way. The quantum signals that carry the key information are individual photons or photon pairs. Detecting and generating single photons with sufficient fidelity requires specialized optical hardware: single-photon detectors, precisely tuned lasers, interferometers, and sometimes exotic cooling equipment. None of this lives in your laptop or your phone or your server rack at the moment you decide you want QKD. It must be physically installed, calibrated, and maintained.

Diagram 10.3: The hardware gap: software update vs. QKD installation

The fiber requirement adds another layer of constraint. QKD typically uses photons traveling through optical fiber or through free-space line-of-sight paths. Standard telecom fiber carries multiple data channels simultaneously, using different wavelengths of light; this is called wavelength-division multiplexing. The intense classical light signals in a multiplexed fiber can interfere with the delicate single-photon quantum signals. Some QKD deployments require entirely dedicated fiber links, meaning fiber that carries no classical traffic. Dedicated dark fiber is expensive to lease and install, especially over long distances.

Research has made progress in running QKD and classical signals together over the same fiber using carefully separated wavelengths and power levels, and this remains an active and genuinely promising area. But it remains more technically demanding than purely classical fiber links, and it is not yet the universal standard. For organizations considering QKD, the fiber situation is often the most immediately sobering practical constraint.

Satellite QKD avoids fiber entirely by using free-space optical links, and experiments have demonstrated that this approach can work over distances of thousands of kilometers. But satellite links require ground stations with precision telescopes, clear skies, and low atmospheric turbulence. A satellite QKD link that works beautifully on a clear night in a mountain observatory faces much harder conditions in a cloudy coastal city. Weather is not an abstraction; it is a real operational variable for any free-space quantum channel.

The hardware constraint has a deeper implication as well. Because QKD requires specialized equipment, it requires specialized supply chains, trained technicians, and specialized repair procedures. An organization that deploys QKD must plan for the hardware lifecycle: procurement, installation, calibration, maintenance, and eventual replacement. This is not the same as managing software. It is closer to managing specialized laboratory equipment or telecommunications infrastructure. For large governments and financial institutions, this may be entirely feasible. For a small hospital or a regional utility, it represents a real barrier.

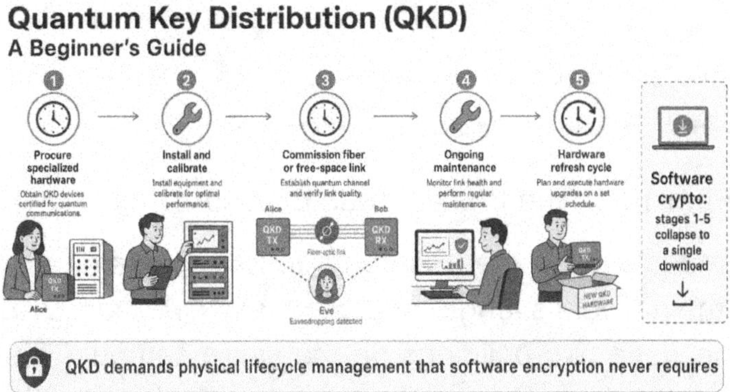

Diagram 10.4: The QKD hardware lifecycle

The Trusted-Node Problem, Revisited

Chapter 7 explained the trusted-node solution in full: because a quantum signal cannot be amplified, long-distance QKD breaks the path into short segments and relays the key from node to node, with the key briefly existing in the clear at every relay node. Here the point is simply what that means for security. Each relay is a real building in a real jurisdiction, staffed by real people, and anyone who compromises one - through hacking, physical access, a corrupt insider, or a coercive order - can read every key that transits it. A continental network may chain dozens or hundreds of such nodes, and each must be secured to the same standard as the endpoints themselves.

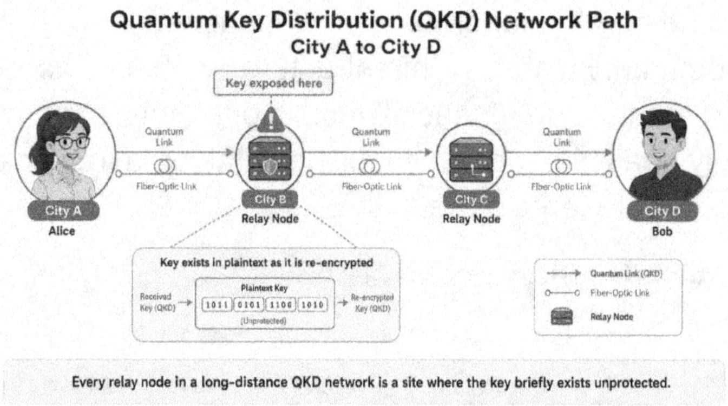

Diagram 10.5: The trusted-node exposure point

The magnitude of the trusted-node risk scales directly with the network's length and complexity. A short QKD link between two buildings on the same campus, with no relay nodes, inherits almost none of this risk. The trusted-node problem does not apply. But a continental-scale network that relies on many relay points reintroduces, in a slightly different form, exactly the kind of insider and physical-access risk that motivated QKD's development in the first place.

Researchers are working hard on solutions. Quantum repeaters, which would allow the relay step to happen without the key ever existing in the clear, are the long-term technical goal. A true quantum repeater would preserve quantum entanglement over long distances by storing and re-emitting quantum states without measuring them, thereby avoiding collapse of the key. But quantum repeaters capable of operating at the scale and reliability required for real networks do not yet exist outside controlled laboratory experiments. The gap between laboratory demonstration and deployed repeater infrastructure remains measured in years, perhaps decades.

Satellite QKD offers a partial workaround for terrestrial nodes. A satellite can distribute entangled photon pairs to ground stations thousands of kilometers apart, allowing those stations to establish a shared key without any intermediate relay node on the ground. Experiments have shown this works over intercontinental distances. But satellite links bring their own limitations: limited transmission windows, weather sensitivity, and the political and economic complexity of satellite infrastructure. And a satellite itself, if compromised, becomes a trusted node in space.

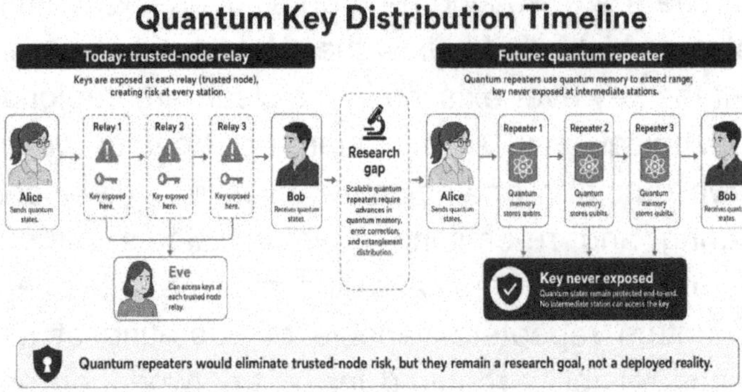

Diagram 10.6: Quantum repeater as the long-term solution

The honest summary of the trusted-node situation is this: long-distance QKD in its current deployed form is not end-to-end physics-secure. It is end-to-end secure only if every relay node is operationally secure, meaning secure by locks, guards, and institutional controls rather than by quantum physics. This does not make current QKD useless; many critical systems already rely on operationally secured nodes. But it does mean that the phrase "quantum-secure communication" should be used with precision. It applies cleanly to short-range, no-relay QKD links. For long-distance networks, the qualifier "pending trusted-node security" belongs in any honest description.

Security Depends on Implementation

The theoretical proof at the heart of QKD is elegant: measuring a quantum state disturbs it, so any eavesdropper leaves detectable traces, so a clean quantum channel proves the key is private. This proof holds for ideal devices operating according to perfect physical models.

Real devices are not ideal. And the gap between the theoretical model and the physical hardware is where implementation attacks live.

Consider the single-photon detectors used in many QKD systems. These detectors are supposed to click once when a single photon arrives. In the theoretical model, they behave perfectly. In the real world, they can be "blinded" by flooding them with a continuous bright light source. When a detector is blinded, it stops responding to individual photons in the normal way. Instead, it enters a regime in which it can be triggered at will by an attacker who sends precisely timed bright pulses superimposed on the blinding signal. This is called a detector-blinding attack, and it was experimentally demonstrated against real commercial QKD systems under laboratory conditions. The physics of the theoretical proof was not broken; the attack exploited the gap between the ideal detector assumed in the proof and the actual detector on the bench.

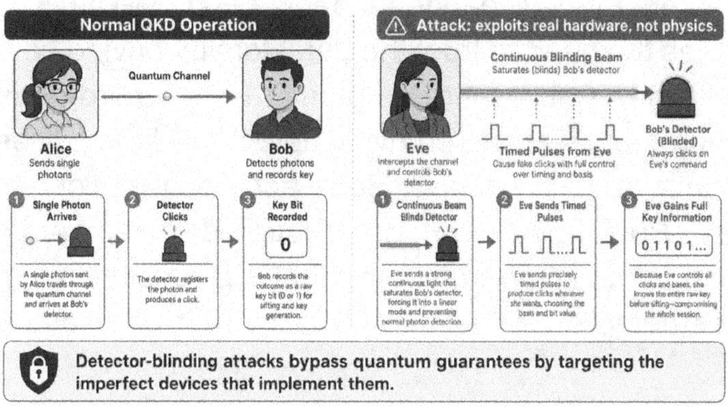

Diagram 10.7: The detector-blinding attack

The detector-blinding attack is the most famous, but it is not the only class of implementation attack. Researchers have demonstrated attacks based on imperfect photon sources that sometimes emit more than one photon per pulse (called photon-number-splitting attacks, discussed in earlier chapters), timing side-channels that leak information through the precise arrival times of photons, and phase-remapping attacks that exploit slight imperfections in how interferometers are tuned. Each of these attacks follows the same pattern: the theoretical security proof is intact, but the real hardware does not perfectly match the assumptions on which the proof is built.

The quantum security community has developed responses to these attacks. Device-independent QKD (DI-QKD) aims to remove the hardware assumption entirely by basing security on measurement statistics that are impossible to fake even with arbitrarily bad detectors, provided the detectors cannot communicate with the outside world during measurement. Measurement-device-independent QKD (MDI-QKD) removes the need to trust the detectors by having both Alice and Bob send photons to a third-party measurement station, removing detector vulnerabilities from the security model. Both approaches work in theory. Both have been demonstrated in laboratory settings. Both come with costs: DI-QKD requires much higher detection rates than current hardware easily achieves, and MDI-QKD requires careful photon engineering at both ends.

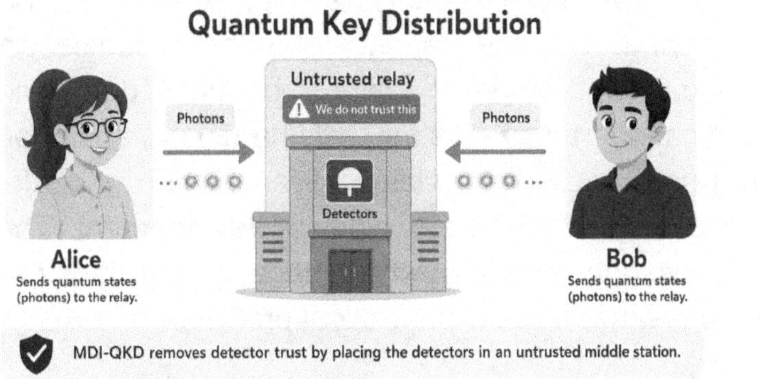

Diagram 10.8: MDI-QKD as a response to detector attacks

There is also a broader lesson here that applies to all security systems, not just QKD. No security protocol, quantum or classical, can be more secure than the hardware and software that implement it. The history of classical cryptography is full of examples where a faulty implementation defeated a mathematically perfect algorithm: poor random number generators, memory timing leaks, padding oracle attacks, and dozens of others. QKD inherits this general truth. The quantum channel may be physics-protected, but the device that generates the photons, the detector that receives them, the classical processing unit that reads the detector output, and the network interface that passes data to the application all sit outside the quantum protection boundary. An attacker who can compromise any of those components can potentially compromise the key.

This is not a reason to give up on QKD. It is a reason to demand that QKD systems be accompanied by rigorous hardware security testing, certification, and monitoring. Several national standards bodies and testing laboratories have begun developing certification frameworks for QKD hardware, analogous to those applied to classical encryption modules. These frameworks are a positive development. They acknowledge that real-world security requires real-world testing, not just theoretical proofs.

Cost, Distance, and Denial of Service

Let us talk about three practical constraints that receive less attention than the conceptual issues but matter enormously for anyone thinking about deploying QKD.

The first is cost. QKD hardware is expensive. The single-photon detectors, specialized lasers, precise optical alignment hardware, and control electronics are all specialized items that do not benefit from the mass-production economies that make classical networking hardware cheap. A QKD link capable of generating useful key rates over a few tens of kilometers costs orders of magnitude more than an equivalent classical encryption device covering the same distance. For organizations with an unrestricted security equipment budget, this may be acceptable. For most organizations, it is a significant barrier.

The cost situation is improving. As QKD has moved from laboratories toward commercial products, more vendors have entered the market, driving some price reduction. Chip-scale integration, where the optical components of a QKD transmitter or receiver are miniaturized onto a photonic integrated circuit, is a promising direction for further cost reduction. But QKD has not yet reached the point where it is economically competitive with classical encryption for typical use cases. It remains, for now, a technology for high-stakes, high-budget applications.

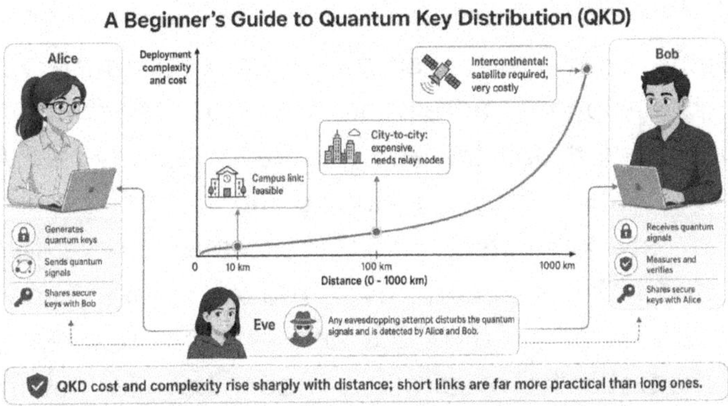

Diagram 10.9: Cost and distance constraints on QKD

The second practical limit is range. In optical fiber, photons are absorbed and scattered as they travel. Classical optical networks overcome this by using amplifiers, devices that boost the optical signal every 50 to 100 kilometers along the path. But quantum states cannot be amplified without being measured, and measuring them collapses the quantum property that makes QKD work. This is the no-cloning theorem at work in the real world: you cannot copy or amplify a quantum state without disturbing it.

The practical consequence is that a single QKD link without relay nodes is typically limited to roughly 100-

200 kilometers over standard fiber, depending on detector quality and the specific protocol used. Research using extremely quiet detectors cooled to near-absolute zero and specialized fiber has extended this to several hundred kilometers, but not to arbitrary distances. For networks that must span continents, relay nodes or satellites are the only current options, each of which brings its own complications as described earlier.

Key generation rate is also a function of distance. As the link gets longer, more photons are lost, fewer arrive at the detector, and the key generation rate drops. A QKD link that generates millions of key bits per second over a few kilometers might generate only a few thousand bits per second over 100 kilometers. Since the one-time pad (OTP), the only encryption scheme that theoretically requires a fresh key for every message, consumes key material at the same rate as data, low key generation rates can become a practical bottleneck. QKD works well for applications that require modest amounts of highly protected key material; it is harder to use for applications that need to encrypt high-bandwidth data streams continuously.

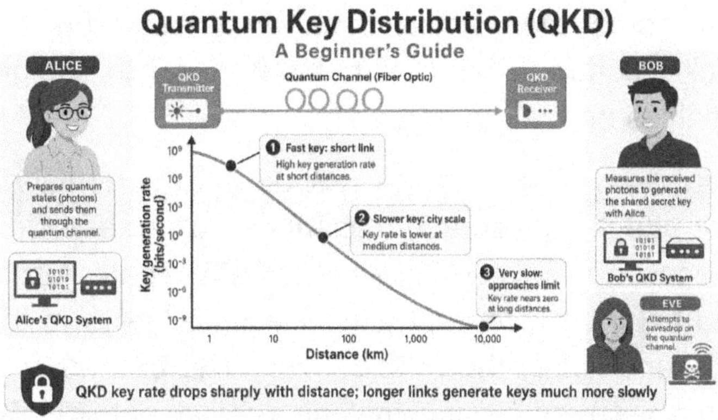

Diagram 10.10: Key rate vs. distance tradeoff

The third practical limit is denial-of-service, which rarely appears in QKD marketing materials. Classical encryption, once keys are established, works even if an attacker is actively interfering with the communication channel. You can detect the interference, but you may still be able to receive and decrypt messages through redundant paths. QKD, by contrast, requires a quantum channel that is quiet and high-quality. If an attacker blocks, cuts, or floods the quantum channel with noise, the QKD session cannot complete. No key is generated. No communication can proceed using QKD-derived keys.

An attacker who cannot crack the key can still stop it from forming. This is a fundamentally different threat model from the one classical encryption faces. It means that QKD systems require physical protection of their quantum channels in addition to all the other security measures. A fiber link used for QKD is not just a data pipe; it is a security-critical physical asset. Cutting that fiber is not just a service interruption; it is a way to force the parties back onto less-protected fallback communication methods, which might themselves be the real target.

The good news is that the response to a denial-of-service attack on a quantum channel is clear: establish the necessary keys over an alternative path, using post-quantum cryptography or pre-shared keys in the interim, and repair the physical link. The bad news is that this requires having that alternative path ready, which means QKD cannot be the only security layer in any resilient system design.

What the Skeptics Say

It is important to engage seriously with the critics of QKD, because several major national security agencies in multiple countries have issued formal assessments recommending against QKD for their most sensitive uses. These are not uninformed critics. They have access to classified threat assessments, deep technical expertise, and operational experience with deploying cryptography at scale. Their concerns deserve more than a dismissive paragraph.

The core of the skeptical argument can be summarized in four points.

The first point concerns authentication. As covered earlier in this chapter, QKD does not solve the authentication problem. Any practical QKD deployment requires an additional authentication mechanism, usually one based on classical cryptography. But if you already have strong classical cryptography (including post-quantum cryptography) running alongside your QKD system to handle authentication, the marginal security benefit of QKD narrows. You are adding significant hardware cost and operational complexity to a system that already has a robust cryptographic foundation.

The second point concerns hardware complexity. The more complex a security system, the larger its attack surface. A QKD system introduces new hardware, software, calibration procedures, supply chain dependencies, and maintenance requirements. Each of these is a potential vulnerability. Post-quantum cryptography, by contrast, is a software update to existing infrastructure. It inherits the well-tested security properties of that infrastructure and does not add new physical attack surfaces. The skeptics argue that, from an operational security standpoint, simplicity is a virtue.

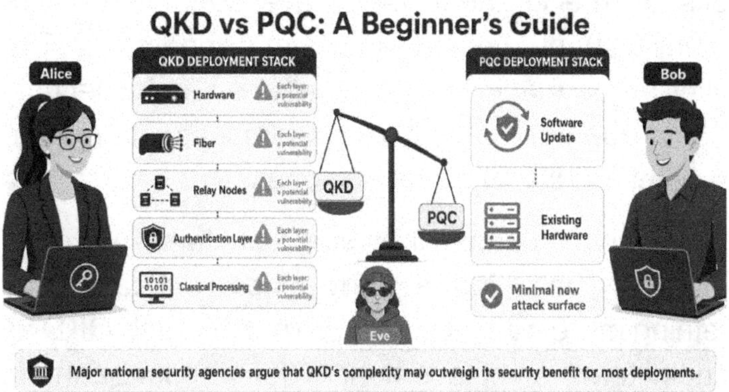

Diagram 10.11: The skeptics' case: complexity vs. marginal gain

The third point is operational. QKD requires specialized expertise that most organizations lack. Training staff, sourcing hardware, and maintaining quantum optical systems are not tasks a typical IT department can pick up quickly. Post-quantum cryptography, once standardized and implemented in libraries and protocols, can be deployed by standard software engineering teams. For large-scale national deployments, the human capital constraint is real.

The fourth point is harder. The security proof for QKD assumes a specific threat model: an eavesdropper who has access to the quantum channel but cannot compromise the endpoint hardware. If an adversary can compromise the hardware, or the software that processes the key material, or the people who operate the system, then the quantum channel's security is irrelevant. And in practice, the most devastating attacks on cryptographic infrastructure have not been attacks on the channel; they have been attacks on endpoints, supply chains, and human operators. The skeptics argue that investing in better endpoint security and post-quantum algorithms addresses a broader, more common threat than QKD does.

None of this means the skeptics think QKD is worthless. Their formal advisories typically say something more nuanced: they do not endorse QKD as a near-term replacement for post-quantum cryptography, and they are concerned about organizations deploying QKD in ways that create a false sense of security while neglecting other vulnerabilities. That is a reasonable concern, stated carefully.

What the Adopters Say

The nations and institutions that are actively deploying QKD have heard the skeptical arguments and are not persuaded that they add up to a reason to stop. Their counter-arguments are worth understanding in equal detail.

The most fundamental counterargument concerns threat timelines. Post-quantum cryptography is based on mathematical problems that current computers cannot solve efficiently. Those mathematical problems are considered difficult, and the standardization process for post-quantum algorithms is rigorous. But "believed to be hard" is not the same as "proven to be hard." The history of classical cryptography includes several algorithms that were believed to be secure and were later broken, not always by the mechanism originally feared, but broken nonetheless. If a future mathematical breakthrough weakens a post-quantum algorithm, there may be very little warning before the attack is operational.

QKD's guarantee does not have this conditional structure. The security of a QKD session does not depend on any mathematical assumption. It depends on quantum mechanics, which is the most precisely tested physical theory in human history. No experiment has ever found a violation of quantum mechanics. The adopters argue that this is a qualitatively different kind of security guarantee, one worth paying for in additional hardware cost and operational complexity, at least for the most sensitive communications.

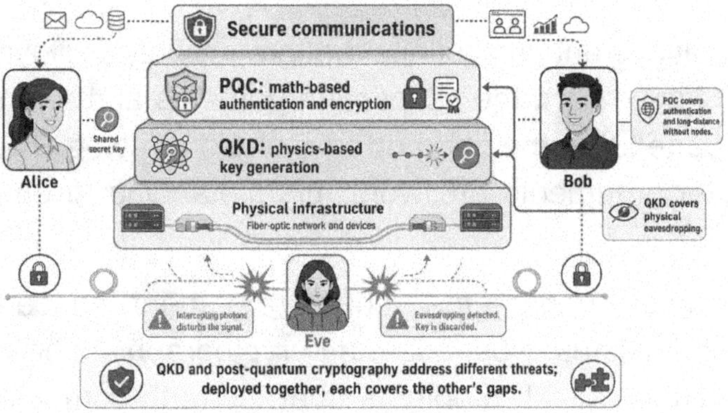

Diagram 10.12: QKD and PQC as complementary layers

The second counter-argument concerns the "harvest now, decrypt later" threat, also sometimes called "store now, decrypt later". This is the scenario where an adversary collects encrypted data today, stores it, and waits until quantum computers are powerful enough to break current classical encryption, then decrypts it at their leisure. Governments, financial institutions, and healthcare systems all generate data today that must remain confidential for decades. A state actor with long planning horizons could be harvesting such data right now. QKD protects against this threat in a way that is independent of future computational advances. Data encrypted under a QKD-generated key cannot be broken by future quantum computers, not because the math is hard, but because the physics prevents the key from ever having been intercepted.

The adopters also argue that the authentication problem, real as it is, is manageable. Many high-stakes QKD deployments are between fixed, known endpoints: two data centers belonging to the same organization, a bank and its regional processing hub, a government ministry and its secure facility. In these cases, the initial authentication ceremony, sharing a short pre-shared key by a trusted courier or secure in-person exchange, is a one-time setup cost that is entirely practical. It is not the same as the open-internet problem of authenticating strangers.

The trusted-node problem also looks different through the adopters' lens. For national infrastructure operators, managing and securing a network of relay stations is a common operational requirement. They already manage physical infrastructure with strong security requirements: power substations, water treatment facilities, data centers. Adding a set of rigorously secured QKD relay nodes to their existing physical security apparatus is an extension of something they already do, not a completely novel challenge.

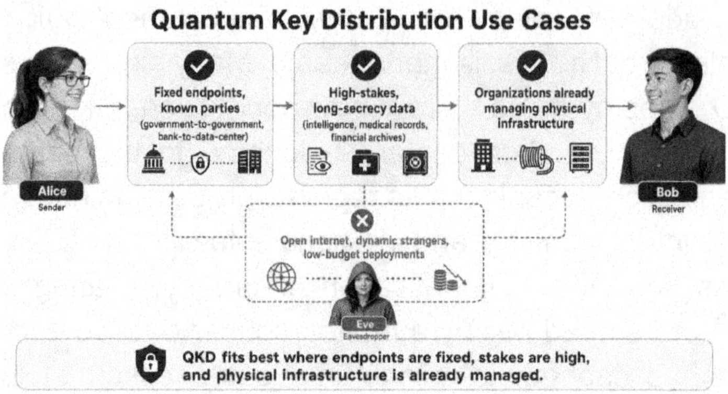

Diagram 10.13: The adopters' case: where QKD fits best

The most compelling argument from the adopters is complementarity. QKD and post-quantum cryptography are not rivals competing for the same role. They address different threat scenarios. Post-quantum cryptography is a software-deployable defense against the possibility that quantum computers could break current math-based algorithms. QKD is a hardware-based defense against eavesdropping on the quantum channel itself, with a guarantee that does not depend on mathematical assumptions. A security architecture that uses both is more resilient than one that uses either alone. The QKD layer handles key generation with physics-based security. The PQC layer handles authentication and provides a fallback encryption layer with math-based security. Neither layer needs to be perfect, because the other layer covers its gaps.

This is the frame that most serious QKD researchers and many deploying nations have converged on: not

"QKD vs. PQC" but "QKD plus PQC." The debate is not which technology to choose; it is whether the total cost and complexity of deploying both is justified by the security benefit in a given application. For most ordinary applications, it probably is not. For the most sensitive communications of the most critical organizations, it increasingly is.

QKD and Post-Quantum Cryptography Are Partners

It is worth spending a few more paragraphs on the theme of complementarity, because it is the most important reframing this chapter offers. Much of the public discussion about QKD frames it as competing with post-quantum cryptography, as if choosing one means rejecting the other. This framing is misleading, and it leads both enthusiasts and skeptics astray.

Post-quantum cryptography solves a specific problem: it replaces existing classical algorithms (primarily those based on the difficulty of factoring large numbers or solving discrete logarithm problems) with new algorithms that are believed to be resistant to attacks by quantum computers. The standardization of PQC algorithms by national standards bodies is a major positive development. These algorithms will run on existing hardware, integrate into existing protocols, and protect data against quantum computers that may emerge in the coming years or decades. They should be deployed broadly and relatively soon.

But PQC does not protect against an adversary who can listen to the quantum channel itself. If an adversary can capture and store the actual photons in your fiber today, they could, in principle, use a future quantum computer to extract information from those photons in ways that are not yet fully characterized. This is a frontier research question, not a settled threat. But it illustrates that PQC and QKD are protecting against different attack surfaces.

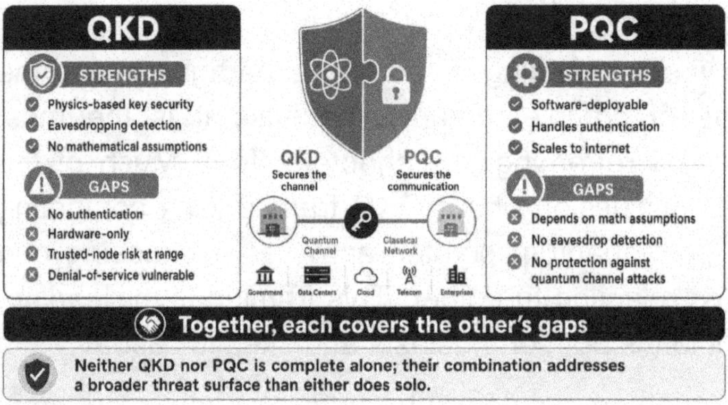

Diagram 10.14: The full honest picture: strengths and gaps

QKD provides something PQC does not: a real-time, physics-based alarm that tells you whether someone is listening to your channel right now. If the error rate on your quantum channel is too high, you know someone is there. You do not need to wait for a mathematical advance to reveal the attack. The detection is built into the channel's physics. This is a genuinely unique property that no amount of improved mathematics will replicate.

PQC provides something QKD does not: the ability to authenticate parties who have never met, to operate at internet scale, to run as software on commodity hardware, and to function over any communication channel regardless of its physical properties. These properties are essential for securing the general internet, for enabling new organizations to communicate securely without physical setup ceremonies, and for protecting a world where the number of secure connections is measured in billions.

The intellectually honest conclusion is that QKD is a powerful, specialized tool with real limitations, and that PQC is a powerful, broad-coverage tool with different limitations. An organization with the resources and the mandate to deploy both gets a security architecture that is stronger than either alone. An organization that must choose, because of cost or operational constraints, should understand clearly what each tool offers and what it leaves unprotected. That is the honest picture this chapter has been built to provide.

Key Concepts Checklist

• I can explain why QKD does not solve authentication on its own, and describe how a man-in-the-middle attack can exploit this gap.

• I understand why QKD cannot be delivered as software, and what dedicated hardware and fiber requirements mean operationally.

• I can describe the trusted-node weakness as a first-class limitation that affects long-distance QKD networks today.

• I understand that implementation attacks have defeated real QKD hardware by exploiting the gap between ideal device models and physical reality.

• I know how cost, range limitations, and denial-of-service vulnerability constrain QKD deployments in practice.

• I can summarize the skeptics' position fairly, including their concerns about authentication, complexity, and operational cost.

• I can describe why many deployers see QKD and post-quantum cryptography as complementary rather than competing.

Questions to Explore

• If QKD cannot prove who is on the other end of the channel, how serious is that limitation for a government agency deploying QKD between two of its own fixed facilities? How does your answer change if those facilities are connected by a network that crosses a third country's territory?

• Why might a national security agency reach a different conclusion about QKD than a national telecom operator or a large financial institution? What does each organization optimize for, and how do those priorities shape the QKD calculus?

• A denial-of-service attack on a quantum channel cannot break the key, but it can stop the key from forming. Is this a fatal flaw in QKD's security model, or a manageable operational risk? What would a resilient system design look like if it needed to survive such an attack?

• After reading this chapter, has your assessment of QKD's value gone up, down, or become more precise? What specific piece of information most changed your thinking?

Try This / Reflect

Find one public document that is skeptical of QKD (a government advisory or a security researcher's analysis is a good starting point) and one that is enthusiastic about it (a research paper from a QKD research group or a national QKD program report). Read both and list the specific claims each makes. Then sort every claim into one of three categories: claims about the underlying physics, claims about hardware and implementation, or claims about authentication and operational factors. You will very likely find that the two sides are not disagreeing about the physics at all. The enthusiasts and the skeptics agree on what quantum mechanics does. Their disagreement lives almost entirely in the hardware, authentication, and operational layers. Seeing that clearly is the most useful outcome of this exercise because it shows that the debate is not between people who understand QKD and those who do not. It is between people who weight hardware cost and operational complexity differently, and that is a question with room for reasonable disagreement.

Glossary Terms

• **Authentication -** The process of verifying the identity of the party you are communicating with, distinct from encryption, which protects the content of the communication but does not establish who the sender or receiver is.

• **Trusted-node risk -** The vulnerability introduced by relay stations in a long-distance QKD network, where the key must briefly exist in the clear at each relay point, making each relay a potential point of compromise.

• **Implementation attack -** A method of defeating a security system not by breaking its underlying theory but by exploiting imperfections in the hardware or software that implements it; the detector-blinding attack on QKD systems is a well-known example.

• **Denial of service -** In the QKD context, an attack in which an adversary disrupts or blocks the quantum channel to prevent key generation, without attempting to read the key; a type of attack that classical encryption channels are less vulnerable to.

• **Post-quantum cryptography (PQC) -** Mathematical encryption algorithms designed to resist attacks by future quantum computers, deployable as software on existing hardware; addresses a different set of threats than QKD and is best understood as complementary to it rather than competing.

• **Complementary security -** The principle that two security tools with different strengths and weaknesses can be deployed together so that each covers the gaps of the other; in QKD deployment, this describes the relationship between QKD and post-quantum cryptography.

Key Takeaways

• QKD secures the quantum channel against eavesdropping but cannot authenticate who is at either end; authentication must come from an additional mechanism.

• QKD requires specialized hardware and usually dedicated fiber or line-of-sight links; it cannot be deployed as software or a cloud service.

• Long-distance QKD networks depend on trusted relay nodes where keys exist briefly in the clear, reintroducing physical and insider risks at each relay point.

• Real QKD systems have been defeated by implementation attacks that target hardware imperfections rather than quantum physics; real-world security requires hardware testing and certification, not just theoretical proofs.

• Major national security agencies are cautious about QKD for operational reasons, particularly around authentication, complexity, and cost; the most constructive response to their cautions is to deploy QKD alongside post-quantum cryptography rather than instead of it.

11 QKD vs Post-Quantum Cryptography

Overview

The quantum threat explored earlier in this book has produced two very different defensive responses. The first is quantum key distribution (QKD), which you have now studied at length: a method that uses particles of light to share cryptographic keys in a way governed by the laws of physics rather than the limits of computing power. The second is post-quantum cryptography (PQC): an approach that keeps the familiar software-based architecture of today's encryption but replaces the underlying math with new problems believed to be too hard for even a powerful quantum computer to crack quickly. These two responses are not rivals in the ordinary sense. They are complementary tools, each with its own strengths, costs, and risks, designed for different parts of the same problem.

This chapter gives PQC the fair and careful treatment it deserves, then places it side by side with QKD so you can see exactly where they agree, where they diverge, and why the security community is increasingly convinced that the smartest path forward is to use both. PQC can be deployed as a software update to virtually any device already connected to a network, and it solves a problem QKD cannot solve on its own: verifying who you are actually talking to. QKD, in return, offers a security guarantee that does not rest on any mathematical assumption, which is a fundamentally different kind of assurance. Neither technology is without risk, and this chapter is honest about both. The goal is to leave you with a clear mental model so that the next time you encounter a vendor pitch, a news headline, or a policy debate that frames these two as competitors, you will know exactly what is actually being said.

Opening Scenario

A security manager at a regional hospital network receives two vendor proposals in the same week. The first vendor describes a software package that can be rolled out to every laptop, server, and medical device on the network within days. It promises protection against quantum computers by replacing the hospital's current encryption algorithms with newer, mathematically stronger ones. No new hardware is required. The update costs a fraction of the annual IT budget. The second vendor proposes installing specialized fiber-optic hardware on the single connection between the hospital's main data center and its off-site backup facility, the link that carries the most sensitive patient records in real time. The hardware would generate encryption keys using quantum physics, guaranteeing that no eavesdropper could intercept them without leaving a trace. The cost is high, and the technology only covers that one link.

The manager assumes these are competing bids for the same job. She asks her team to pick one. Her team is not sure how to compare them. The proposals use different language, cite different standards bodies, and make different kinds of promises. One is talking about math; the other is talking about physics. One covers everything; the other covers one cable. Neither vendor has explained why both might be the right answer at the same time.

This chapter is the clear comparison the manager wished the vendors had provided. It explains what post-quantum cryptography is, how it differs from QKD at a fundamental level, where each excels, what risks each carries, and why the strongest approach often layers them together rather than choosing between them.

What This Chapter Will Cover

- **Two roads from the same threat:** how QKD and PQC both respond to the quantum computing challenge but do so in completely different ways.
- **What post-quantum cryptography is:** a plain-language explanation of PQC, the math behind it, and why global standards bodies have already selected these algorithms for wide deployment.
- **The big difference: math vs. physics:** the core contrast between a security guarantee rooted in an unproven mathematical assumption and one rooted in physical law.
- **Deployment: software everywhere vs. hardware somewhere:** why PQC can reach billions of devices while QKD serves specific, high-priority links.

• **Authentication: a decisive advantage for PQC:** why the ability to prove identity matters as much as encrypting content, and why QKD cannot do this on its own.

• **The risk each one carries:** an honest account of where PQC and QKD can each fail or fall short.

• **Why the future is both:** the case for layering QKD and PQC together, which is where the field is genuinely heading.

Two Roads from the Same Threat

Every few decades, a new kind of computational power forces cryptographers to reconsider the foundations of their field. The quantum computing challenge described earlier in this book is the most significant such moment since the invention of public-key cryptography in the 1970s. The algorithms that secure your email, your bank transactions, and your medical records all rely on the difficulty of certain mathematical problems: factoring large numbers, or finding discrete logarithms in enormous number fields. A sufficiently powerful quantum computer, using an algorithm designed specifically for the task, could solve those problems far faster than any classical machine. That threat is not hypothetical. It is a matter of engineering timeline, and the engineering is advancing steadlly.

Faced with that threat, the world's cryptographers and security engineers split into two broad camps. One camp said: change the physics. Instead of relying on mathematical hardness, use the quantum properties of light itself to distribute encryption keys. Any eavesdropper who tries to intercept a quantum signal disturbs it in a detectable way, because quantum mechanics says you cannot measure a quantum state without altering it. That camp built QKD. The other camp said: keep the software approach, but change the math. Find new mathematical problems that even quantum computers cannot crack efficiently, and build encryption algorithms on top of those problems. That camp built PQC.

Both camps were responding to the same danger. Both produced serious, well-funded, internationally recognized technologies. The key insight is that they did not produce the same product. QKD is a hardware system for distributing keys over specific physical links. PQC is a software upgrade for all existing forms of encrypted communication. Understanding that distinction is the entire foundation of this chapter.

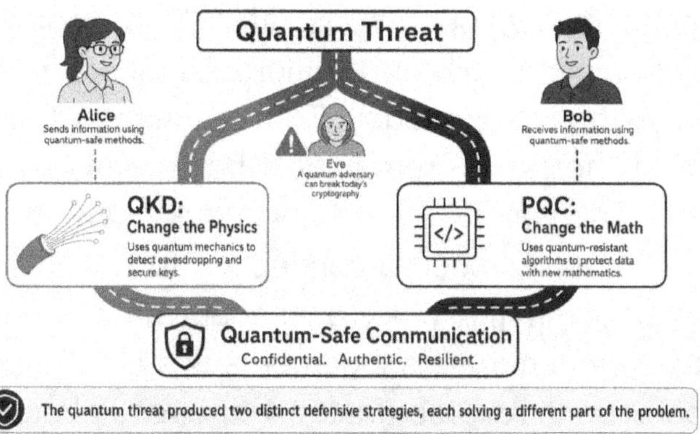

Diagram 11.1: Two roads from one quantum threat

The fact that two camps emerged does not mean they are at war with each other. Most serious security architects today treat QKD and PQC as tools in the same toolkit, not as rivals. But to understand why they belong together, you first have to understand each one on its own terms. We will spend the next several sections doing exactly that, starting with the technology that is less familiar to most readers: post-quantum cryptography.

It is worth pausing to note that the phrase "quantum-safe" appears in marketing for both QKD products and PQC products. This shared label is one of the most common sources of confusion in the field. A product that claims to be quantum-safe might be using physics, might be using new math, or might even be a mixture of the two. By the end of this chapter, you will be able to read those claims critically and ask the right follow-up questions.

What Post-Quantum Cryptography Is

Post-quantum cryptography is, at its core, software-based encryption built on a different foundation than the encryption we use today. To understand what that means, it helps to remember how today's most common encryption algorithms work. When your browser creates a secure connection to a website, it uses an algorithm that relies on the fact that multiplying two large prime numbers together is easy, but running that operation in reverse (factoring the result back into its two primes) is extraordinarily hard. Hard, that is, for a classical computer. A quantum computer running a specific algorithm can factor those numbers efficiently, which means the lock our entire internet infrastructure relies on could eventually be picked.

PQC researchers asked a natural question: are there other mathematical problems that a quantum computer is also bad at solving? The answer, after decades of research, is yes. Several families of mathematical problems appear to resist quantum speedups. Lattice-based problems involve finding short vectors within highly complex geometric structures spanning many hundreds of dimensions. Hash-based problems involve building secure structures from one-way functions that neither classical nor quantum computers can reverse. Code-based problems involve finding errors in large mathematical codes, a task that is hard for quantum algorithms in a way that factoring is not. These are not hypothetical constructs; they are well-studied areas of mathematics with decades of peer-reviewed research behind them.

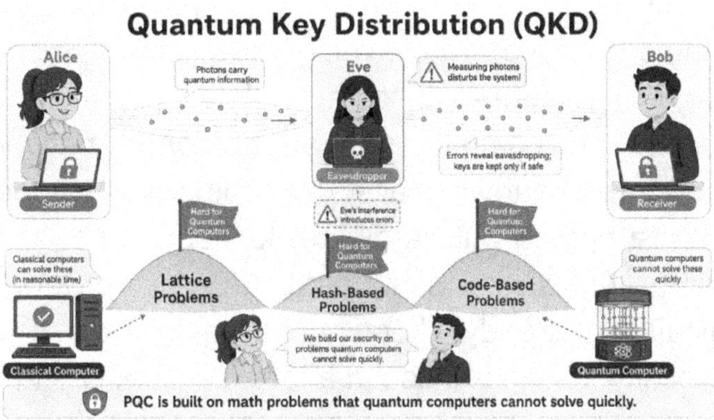

Diagram 11.2: The landscape of post-quantum math problems

The standards body at the center of PQC's development is the National Institute of Standards and Technology (NIST). The same United States agency previously standardized the encryption algorithms now in widespread use. NIST began a formal evaluation process for post-quantum algorithms in 2016. After multiple rounds of public evaluation, cryptanalysis, and international peer review, it selected a set of algorithms for standardization. The first of those standards was finalized in 2024. This is significant: PQC is not a research project waiting to be validated. It is a technology that has passed one of the most rigorous public review processes in the history of cryptography. Agencies, technology companies, and infrastructure operators around the world are already updating their systems.

What makes PQC particularly powerful in practice is that it fits into the existing software ecosystem. The algorithms produce the same kind of output that today's algorithms produce: a key, a signature, an encrypted message. The format is compatible with the same protocols, programming libraries, and network infrastructure already in place. Upgrading to PQC does not require ripping out hardware. It requires updating software, which is something every networked organization already knows how to do. That deployability, as we will see in a later section, is one of PQC's most significant advantages over QKD.

PQC also covers both major functions of cryptography. It can encrypt data, keeping it confidential. It can also create digital signatures, proving that a message came from a specific party and has not been altered. This second function, authentication, will be discussed at length later in this chapter, because it is an area where QKD alone cannot provide a complete solution.

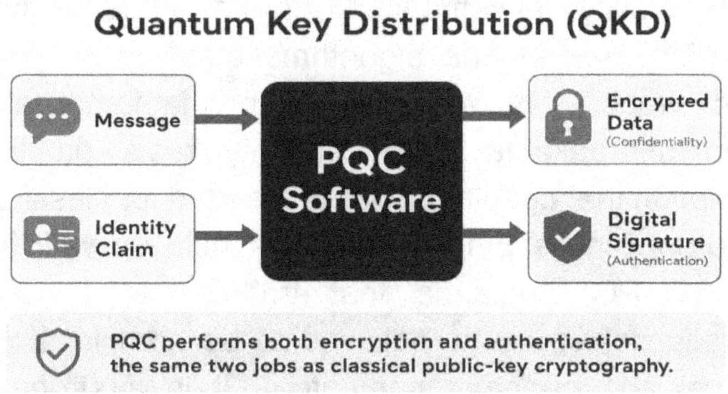

Diagram 11.3: What PQC algorithms actually produce

The Big Difference: Math vs. Physics

At this point you understand the basic mechanics of both QKD and PQC. Now it is time to face the biggest difference between them, because it is not one of performance, cost, or ease of deployment. It is a difference in the kind of promise each technology is making.

PQC makes a mathematical promise. It says: the problems underlying these algorithms are believed to be hard for quantum computers, based on extensive mathematical analysis. No one has found an efficient quantum algorithm to break them. Researchers around the world have been trying for years without success. Thousands of cryptographers have reviewed the algorithms. That body of evidence gives us strong grounds for confidence. But the promise is ultimately an assumption: we believe these problems are hard. We have good reasons to believe it. We cannot prove it in the way a mathematician proves a theorem.

This is actually the same kind of promise that today's encryption makes. The algorithms that secure your bank's website have never been mathematically proven unbreakable. Their security rests on the assumption that certain problems are hard for classical computers, an assumption that has held up well for decades. PQC extends that tradition to a new generation of problems believed to be hard for both classical and quantum computers. It is a stronger assumption than what we have now, but it is still an assumption.

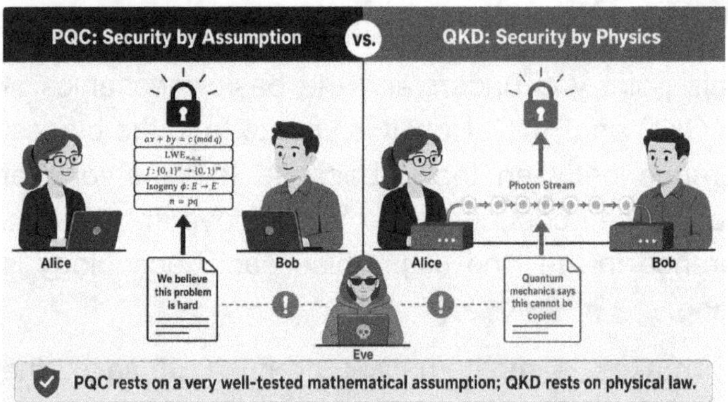

Diagram 11.4: Math assumption vs. physical law

QKD makes a physical promise. It says: the security of the key distribution does not depend on any assumption about computational difficulty. It depends on the behavior of individual photons, which quantum mechanics governs. Quantum mechanics has been tested and confirmed across an enormous range of experimental conditions for nearly a century. When a photon carries a quantum state, measuring it disturbs it. That is not an assumption; it is an observed fact about how the universe works at the smallest scales. An eavesdropper cannot silently intercept a QKD key exchange. She will always leave a detectable trace.

The practical implication of this difference is subtle but important. Mathematical assumptions can turn out to be wrong. History contains examples of encryption algorithms believed secure for years that were later broken, not by quantum computers but by classical mathematics that nobody had thought of yet. A new mathematical insight, even without quantum computing, could in principle threaten PQC. The probability is considered very low, especially for algorithms that have survived years of global scrutiny. But the probability is not zero. Physical law, by contrast, is not subject to revision by a clever mathematician. Nobody is going to discover a new theorem that makes photons stop behaving like photons.

This does not mean QKD is "better." It means QKD provides a different category of guarantee. The analogy is the difference between saying "no one has ever picked this lock in all the years it has been tested" and saying "this lock is made of a material that cannot physically be picked." The first statement might be the result of millions of attempts by the world's best locksmiths failing. It is compelling evidence. The second statement eliminates an entire category of attacks. They are both useful; they are not the same thing.

Diagram 11.5: The locksmiths analogy

Deployment: Software Everywhere vs. Hardware Somewhere

If the math-versus-physics contrast is the most fundamental difference between PQC and QKD, then the deployment contrast is the most practically important one. It explains why most organizations will adopt PQC before QKD and why most of the world's encrypted communications will be protected by PQC for the foreseeable future, even in a world where QKD is also in use.

PQC is software. Software runs on chips. Chips are everywhere: in laptops, smartphones, servers, network routers, embedded controllers in industrial machinery, and smart sensors monitoring a hospital building's temperature. Every one of those devices communicates using encrypted protocols. Every one of those protocols can be updated to use PQC algorithms. The update process is the same one that pushes security patches and operating system upgrades to devices around the world every day. An organization that decides to migrate to PQC does not need to buy new hardware. It needs to work through its software inventory, update its libraries and protocols, compatibility test, and roll the changes out systematically. That is a significant engineering effort, but it is a kind of effort that every IT department already has experience managing.

Diagram 11.6: PQC deployment: software reaching everywhere

QKD is hardware. It requires specialized optical equipment, typically a photon source, a detector, and the fiber-optic or free-space channel connecting them. That hardware must be installed at the endpoints of the specific link you want to protect. It cannot be sent as an over-the-air update. It cannot run on a general-purpose processor. It requires technicians, physical access, calibration, and ongoing maintenance. Every new link you want to protect with QKD requires its own hardware installation.

This is not a criticism. It is simply the nature of the technology. You cannot distribute quantum keys without quantum hardware, just as you cannot bake bread without an oven. The constraint is real, and it means QKD is best suited to specific, high-value links where the physics-based guarantee justifies the cost and complexity. The fiber connection between a central bank's primary data center and its backup facility. The government communications link between two agency headquarters in the same city. The connection inside a hospital between the server holding the most sensitive patient records and the archive where those records are backed up. These are the places where QKD's unique guarantee earns its cost.

The result is a natural division of labor. PQC covers everything. It is the baseline quantum-safe protection that every device and every network link receives as part of the broader migration away from algorithms vulnerable to quantum attack. QKD covers the highest-priority links, the ones where an organization decides that even a very small mathematical risk is unacceptable, or where the confidentiality requirement extends for decades into the future and needs to be immune to any future mathematical breakthrough.

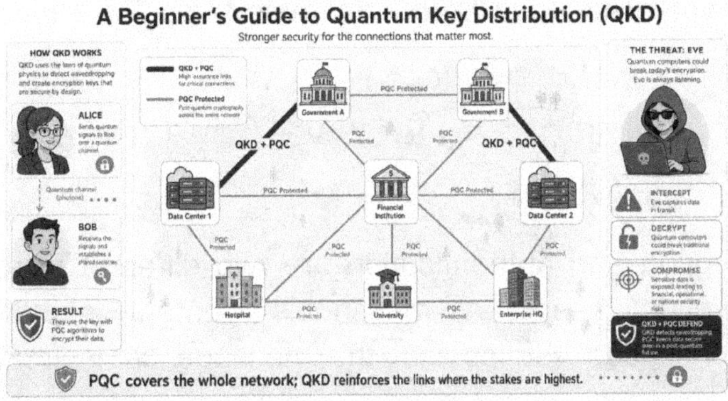

Diagram 11.7: The natural division of labor

Some readers may wonder why QKD cannot simply be made cheaper and more compact over time, eventually reaching the same deployment scale as software. That is a reasonable question, and the answer is nuanced. The physics of QKD does impose genuine constraints that engineering alone cannot remove. Photons are fragile. Individual quantum states cannot be amplified the way classical signals can. Detection requires extremely sensitive equipment operating near absolute zero in some designs. While there has been genuine progress in making QKD hardware smaller and more cost-effective, the constraints it faces are different from those PQC faces. PQC's deployment challenge is organizational: coordinating upgrades across a complex software ecosystem. QKD's deployment challenge is partly physical: the universe imposes certain limits on how quantum information behaves, and those limits shape what the hardware can do. Progress will continue in both areas, but the gap in deployment scale is likely to remain large for a long time.

Authentication: A Decisive Advantage for PQC

Encryption and authentication are two different jobs. Encryption keeps a message's content secret. Authentication verifies that the message came from who you think it came from. Both are essential to secure communication, and they are not the same problem.

Most people encounter authentication in a familiar form. When you log in to your bank's website, the site presents a digital certificate that your browser checks. The certificate proves, through a chain of cryptographic signatures, that you are actually connected to your bank and not to an impostor site that looks identical. Without authentication, encryption is nearly useless for many purposes. An encrypted tunnel to an attacker who has positioned herself in the middle of the connection provides no real security, because everything you type is being read at the other end by the wrong party. This kind of attack, in which a third party pretends to be both sides of a conversation, is called a man-in-the-middle attack, and it is among the most common forms of network intrusion.

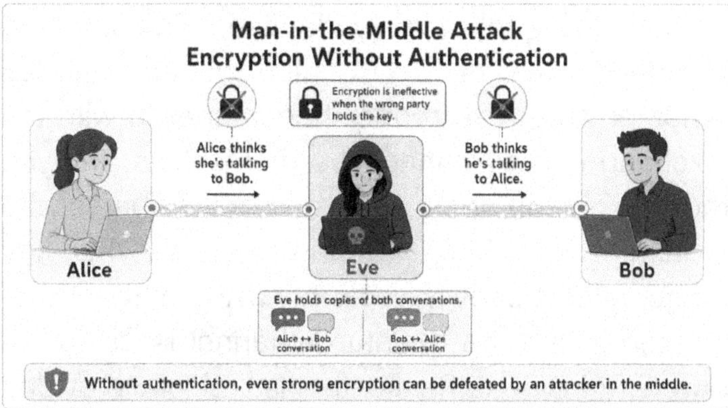

Diagram 11.8: Why authentication matters: the middle attacker

Public-key cryptography, the kind in widespread use today and the kind that PQC replaces and updates, was designed to solve both problems simultaneously. The same mathematical structure that enables encryption also enables digital signatures. A sender can sign a message with a private key she alone holds, producing a signature that anyone can verify with the corresponding public key. The signature proves both the identity of the sender and the integrity of the message. PQC inherits this dual capability. The new algorithms selected by standards bodies produce both encrypted data and digital signatures. A fully PQC-protected network handles authentication the same way the current internet handles it, through certificates, signatures, and public-key infrastructure, just with quantum-resistant math underneath.

QKD, as currently implemented, does not solve authentication on its own. This is one of the most important limitations to understand. QKD is designed to distribute keys, not to verify identities. The standard QKD protocols described in earlier chapters can tell you whether an eavesdropper has interfered with the key exchange. They cannot, by themselves, tell you whether the person at the other end of the fiber is who they claim to be. The two parties in a QKD key exchange must already have some way to authenticate each other before the quantum channel is useful. In practice, that initial authentication is typically provided by classical cryptography, and increasingly it will be provided by PQC. QKD depends on external authentication; it does not provide authentication on its own.

This point has real consequences. Consider a national government that wants to provide quantum-safe security for communications between thousands of officials scattered across dozens of offices in dozens of cities. The officials need to be able to identify each other, send authenticated orders, sign documents, verify approvals, and do all of this over a network where quantum computers might someday be listening. QKD can protect specific fiber links, but it cannot scale to authenticate thousands of individuals. PQC can. A PQC-based certificate infrastructure works the same way as today's internet certificate infrastructure, just with stronger math. Every official gets a digital identity credential. Every message they sign is verifiable. The entire system runs as software on existing hardware.

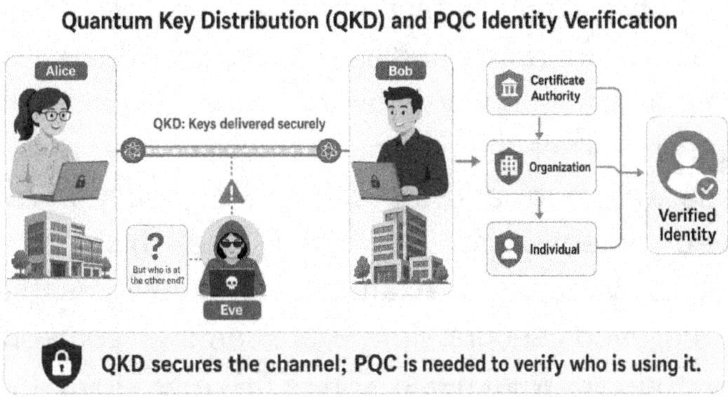

Diagram 11.9: Authentication gap: what QKD alone cannot do

The authentication gap in QKD is not a secret or a scandal. QKD researchers have known about it from the beginning, and the standard response has always been to pair QKD with classical or post-quantum authentication. A system that uses QKD to distribute keys and PQC to authenticate the parties involved is more secure than either technology alone. The key insight is that combining them is not a workaround for a deficiency; it is the architecturally correct approach. The combination uses each technology for the job it does best.

The Risk Each One Carries

An honest evaluation of any security technology includes an honest assessment of what can go wrong. Both PQC and QKD carry real risks. They are different kinds of risk, which is itself an argument for using both together, but neither is risk-free.

PQC's primary risk is mathematical. The algorithms selected by standards bodies were chosen based on the best available analysis by the global cryptography community. That community is large, the review process was long, and the selected algorithms have survived intense scrutiny. But the history of cryptography contains examples of algorithms that were believed secure for years before someone discovered a weakness. The discrete logarithm problems underlying some earlier post-quantum candidates were broken by new classical algorithms before quantum computers even became relevant. A similar development could, in principle, affect some of the currently standardized PQC algorithms. The probability is considered very low, particularly for algorithms in multiple independent families, and the standards bodies have specifically selected algorithms from different mathematical families precisely to reduce the chance that a single breakthrough could undermine everything at once. But the possibility cannot be dismissed entirely.

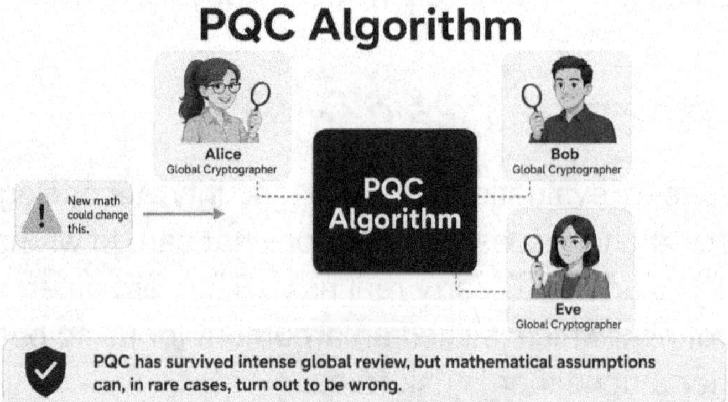

Diagram 11.10: PQC's mathematical risk

There is also a secondary PQC risk worth mentioning: implementation quality. A mathematically sound algorithm can still be undermined by poor software implementation. Side-channel attacks, in which an attacker infers information from the timing or power consumption of a computation rather than breaking the math directly, have successfully compromised cryptographic implementations in the past. Ensuring that PQC algorithms are implemented correctly across billions of devices, libraries, and protocols is a substantial engineering challenge. The field has good tools for this, including formal verification methods and rigorous testing standards, but it requires sustained attention.

QKD's risks were covered in depth in the previous chapter, and it is worth summarizing them here for a complete comparison. The hardware risk is the most straightforward: QKD systems are complex optical devices, and hardware can fail, be misconfigured, or have subtle physical flaws that an attacker can exploit. A photon source that produces photon pairs when it should produce single photons can leak information. A detector that clicks more often for one kind of photon than another can be manipulated. These are not theoretical concerns; actual QKD devices have been demonstrated to be vulnerable to such attacks in laboratory settings, and the engineering community has worked hard to close those gaps. Trust in a QKD system requires trust in the entire physical device chain, not just the underlying protocol.

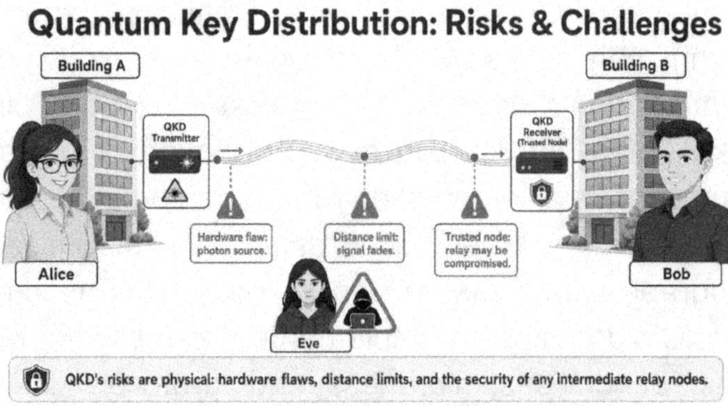

Diagram 11.11: QKD's hardware and distance risks

The distance limit is the second major QKD risk, or rather constraint. Photons carrying quantum states cannot be amplified the way classical signals can. Over long fiber distances, they fade and are lost. Current fiber-based QKD systems operate reliably over distances measured in hundreds of kilometers at best, and that is under good conditions with high-quality fiber. Bridging continental distances requires either satellite-based free-space QKD (which works but introduces its own physical constraints and vulnerabilities) or a network of trusted nodes that relay keys between segments. A trusted node is exactly what it sounds like: a point in the network that must be physically secure and operated by a party you trust. If any trusted node is compromised, the security of the key distribution through that node is compromised. This is a real limitation for large-scale deployments, and it is one that QKD's fundamental physics does not fully resolve.

The risk profiles of PQC and QKD are, notably, different in character. PQC's main risk is that a mathematical breakthrough could weaken the algorithm. QKD's main risks are physical: hardware imperfections, distance constraints, and relay-node security. An organization that uses both technologies faces a situation in which breaking its security requires a simultaneous mathematical breakthrough against PQC and a physical compromise of the QKD hardware or relay infrastructure. That is a much higher bar than either technology sets alone, which is the core argument for combining them.

Why the Future Is Both

The title of this section is not a vague hope for harmony. It is a description of where the field has already arrived. Major national security agencies, standards bodies, and technology companies have not issued guidance saying "choose QKD or PQC." Their guidance says: deploy PQC broadly and immediately, because it covers everything and its standards are ready; consider QKD for high-priority links where physics-based guarantees are worth the cost and complexity. These two recommendations are not in tension. They point in the same direction.

Think of it in terms of the hospital scenario from the opening of this chapter. The security manager should accept both proposals. The PQC software update protects every device on the network, from administrative laptops to patient monitoring systems to email servers. It provides authenticated, encrypted communication for every channel. It is the quantum-safe upgrade that every device needs, and it is available now. The QKD hardware installation on the single most sensitive fiber link adds a second layer of protection specifically for that link. An attacker who wants to compromise data flowing over that link must now defeat both the PQC authentication and the physics-based key distribution. The two systems reinforce each other in a way that neither can achieve alone.

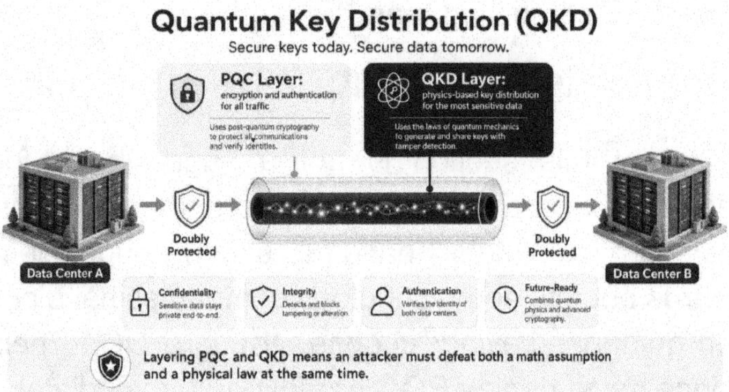

Diagram 11.12: Layering QKD and PQC together

The concept at work here is called **defense-in-depth**. It is a principle borrowed from physical security: never rely on a single line of defense, because every defense has weaknesses. A bank does not rely solely on its front door lock. It uses locks, cameras, guards, vaults, and alarm systems, each layer covering the failure modes of the others. A security architecture that layers

PQC and QKD applies the same logic to both cryptography and QKD. If a mathematical breakthrough were to weaken a PQC algorithm, the QKD layer on the most critical links would still be intact. If a hardware flaw in a QKD device were discovered and exploited, the PQC layer would still be protecting the content of communications. Neither failure mode alone is sufficient to compromise the whole system.

The field is also moving toward formal hybrid protocols: standard communication frameworks that explicitly combine PQC key exchange with QKD key material. The combined key is the product of both processes. Even if one source of key material is compromised, the combined key remains secure as long as the other source is not simultaneously compromised. This is sometimes called a **hybrid key establishment**, and it is the architecturally sound approach that security engineers prefer when the stakes are high enough to justify the complexity.

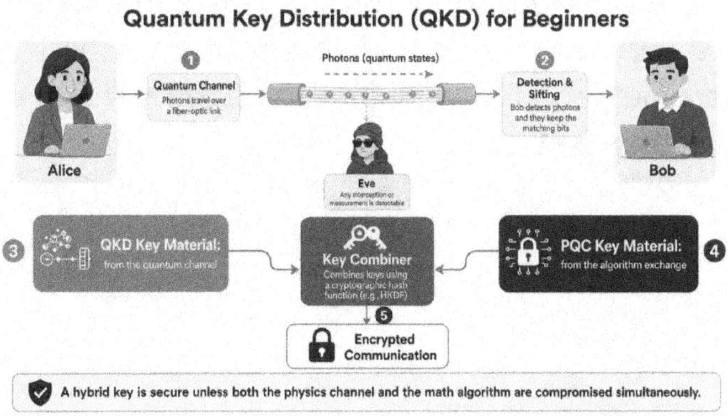

Diagram 11.13: Hybrid key establishment: combining both sources

It is also worth observing that PQC and QKD are not static. Both fields are continuing to develop. PQC researchers are developing additional candidate algorithms and exploring new mathematical structures, building redundancy into the portfolio of available options. QKD researchers are working on satellite-based networks, small-cell quantum repeaters that may eventually eliminate the trusted-node problem, and device-independent protocols that reduce reliance on trusting the hardware itself. The competition between these two fields has, in practice, made both stronger, because each community's advances have prompted the other to raise its standards.

One useful way to think about the current moment is to compare it to the early years of the internet, when organizations were deciding whether to use the newly available firewalls, antivirus software, and encryption together or separately. The organizations that fared best were not the ones that chose one layer and declared it sufficient. They were the ones that layered multiple protections and updated each layer as new threats emerged. The quantum transition will reward the same mindset.

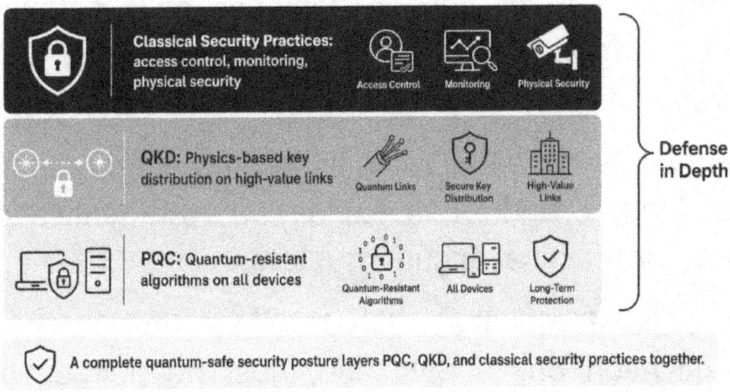

Diagram 11.14: The quantum-safe security stack

The choice between PQC and QKD is a false choice. The real question is not which one to use, but which combination of the two is right for a given organization's links, budget, risk tolerance, and time horizon. For most organizations, the near-term answer is to deploy PQC everywhere as quickly as possible and add QKD to the highest-priority links where the physics-based guarantee and the defense-in-depth benefit justify the investment. That is not a compromise. It is the technically correct answer to the question posed by the quantum threat.

Key Concepts Checklist

• I can explain, in one sentence each, what QKD and post-quantum cryptography (PQC) are and how they differ.

• I understand that PQC's security rests on a mathematical assumption while QKD's rests on physical law, and I can explain why that distinction matters.

• I know why PQC can be deployed as a software update to virtually any networked device, while QKD requires dedicated hardware on specific links.

• I understand that PQC handles both encryption and authentication, and I can explain why authentication matters for practical security systems.

• I understand that QKD does not provide authentication on its own and must be paired with classical or post-quantum authentication for a complete solution.

• I can name the main risk each approach carries: mathematical breakthrough risk for PQC, and hardware plus distance plus trusted-node risk for QKD.

• I can explain the concept of defense in depth and why layering PQC and QKD together provides stronger security than either technology alone.

Questions to Explore

• Which matters more for a specific system you have in mind: a physics-based security guarantee that does not depend on mathematical assumptions, or the ability to deploy protection to every device and authenticate every user? How would you weigh those two priorities against each other?

• PQC rests on new mathematical problems that are less tested than the ones underlying today's encryption. Does that make you more or less comfortable than the hardware risks, distance limits, and trusted-node issues that QKD carries? Is one kind of uncertainty easier to manage than the other?

• In what specific situations would you choose QKD alone, PQC alone, or both together? What features of the situation would drive your decision, and what additional information would you want before committing to a strategy?

• Why might marketing materials blur the line between QKD and PQC, perhaps labeling both simply as "quantum-safe," and what questions would you ask a vendor to clarify which technology they are actually selling?

Try This / Reflect

Spend five minutes with a blank piece of paper before you look back at the chapter. Write down, from memory, the two technologies discussed here and their key properties. For each one, try to write one sentence on what it relies on for security, one sentence on how it gets deployed, one sentence on whether it handles authentication, and one sentence on its main risk. Then write a final sentence describing how the two would work together in practice. When you check your answers against the chapter, pay attention to which properties you remembered clearly and which ones you had to think harder about. The gaps in your recall are the concepts most worth reviewing.

If you have access to a technology news source or a standards body website, search for recent announcements about post-quantum cryptography migration timelines at large organizations. Notice how those announcements describe the relationship between PQC and QKD, if they mention QKD at all. Compare what you find to the framework this chapter has provided, and ask whether the coverage reflects a clear understanding of the two technologies as complementary rather than competing.

Glossary Terms

- **Post-quantum cryptography (PQC) -** Software-based encryption and authentication built on mathematical problems believed to be hard for both classical and quantum computers, providing a drop-in upgrade path for existing encrypted communications.

- **Hard-math assumption -** The foundational belief underlying conventional and post-quantum cryptography that certain mathematical problems cannot be solved quickly; security holds as long as the assumption holds, but it is not a mathematical proof.

- **Physical guarantee -** Security derived directly from the laws of physics rather than from a computational assumption; QKD's eavesdropping detection is an example, resting on the quantum mechanical rule that measuring a quantum state disturbs it.

- **Authentication -** The process of verifying identity in a communication, proving that you are actually communicating with the party you intend to reach; PQC supports authentication through digital signatures while QKD alone does not.

• **Defense in depth** - A security architecture principle in which multiple independent layers of protection are combined so that defeating any single layer is not sufficient to compromise the whole system; applying both PQC and QKD together is a form of defense in depth.

• **Hybrid key establishment** - A protocol that combines key material from two independent sources, such as a QKD channel and a PQC algorithm, so that the resulting combined key remains secure unless both sources are compromised at the same time.

Key Takeaways

• QKD and PQC are two distinct answers to the same quantum computing threat, one changing the physics of key distribution and the other changing the mathematics of encryption algorithms.

• PQC's security rests on mathematical problems believed hard for quantum computers; QKD's security rests on physical law, making them different in the kind of guarantee they offer, not just in degree.

• PQC can be deployed as a software update across virtually any networked device and solves both encryption and authentication; QKD requires dedicated hardware on specific links and depends on external authentication to be complete.

• Each technology carries real risk: PQC faces the possibility of a mathematical breakthrough, while QKD faces hardware imperfections, distance limits, and the need for trusted relay nodes on long paths.

- The strongest approach layers both technologies together, using PQC for ubiquitous authenticated protection across the entire network and QKD for physics-based key distribution on the highest-priority links, achieving a defense in depth that neither technology can provide alone.

12 The Future of Quantum Security

Overview

This final chapter looks honestly at the road ahead. Quantum key distribution (QKD) has traveled a remarkable distance in a short time, moving from a theoretical curiosity in a physics paper to metropolitan fiber networks, satellite links, and government trials spanning thousands of kilometers. But a stubborn limitation still shadows every long-haul deployment: the trusted-node problem. Every relay point in today's long-distance QKD network must be physically secured, because the key is briefly reconstructed there in ordinary digital form. Quantum repeaters are the technology designed to close that gap, and this chapter explains what they would do, how they work conceptually, and why the field is so excited about them while being clear that practical repeaters remain a few steps away from real-world deployment.

Looking further out, the chapter introduces the idea of a quantum internet. This future network would carry quantum information itself rather than just using quantum physics to protect classical keys. It treats that vision with the same calibrated honesty the rest of the book has applied to every topic: genuinely exciting, backed by serious research programs and real investment worldwide, but separated from today's reality by engineering challenges that will take years to overcome. By the end of this chapter you will have a map for evaluating future claims, a set of questions to ask when a headline announces a quantum breakthrough, and the vocabulary to keep learning on your own long after you close this book.

Opening Scenario

Imagine a cabinet in a university physics lab. It looks like a full-sized refrigerator, bristling with optical fiber connections, laser diodes, and a small viewport through which you can see a crystal the size of your thumbnail suspended in a near-vacuum. That crystal is a quantum memory. Its job is deceptively simple to describe and maddeningly difficult to build: accept the fragile quantum state of a single photon arriving from one fiber, hold that state briefly without reading it, and then release the state into a second fiber that connects to the next section of the network. At no point does anything classical happen. There is no measurement, no digital record, no place where a spy could intercept a readable signal.

Right now, that device works for a few milliseconds, in a room kept colder than outer space, operated by a team of physicists who coax it along daily. It is not a product. It is a proof that the idea is physically possible. But every piece of technology you use today, from the smartphone in your pocket to the fiber under the ocean, started as something finicky and expensive in somebody's lab. The question is not whether quantum memories will ever leave the laboratory; the question is how long the journey will take and what the world will look like when they arrive.

This chapter is about that journey. It ties together everything from the previous eleven chapters into a forward-looking picture, honest about today's limits, clear about what progress looks like, and grounded enough to help you navigate the steady stream of quantum headlines that will keep coming long after this book is published.

What This Chapter Will Cover

• The trusted-node problem, revisited one final time, to set up why quantum repeaters matter so much

• What a quantum repeater would actually do, using the same plain analogies that guided the rest of the book

• Quantum memory and entanglement links, the two supporting ideas that make repeaters possible

• Entanglement-based QKD, a family of protocols that uses entangled photon pairs to reduce further the need to trust hardware

• The quantum internet vision, what it would enable, and why it captures the imagination of researchers around the world

• A sober scorecard: what is working today, what is in advanced trials, and what is still a research goal

• How to keep learning and stay grounded, including how to evaluate quantum-security claims and use the appendices in this book as your next steps

The Trusted-Node Problem, One Last Time

You first met the trusted-node problem in Chapter 7, when the discussion turned to building QKD networks that stretch beyond the reach of a single fiber link. A single fiber can carry quantum keys securely for roughly 100 to 150 kilometers before photon loss makes the signal too faint. To go farther, networks use relay stations. At each relay, a key is received from one link and a new key is distributed on the next. The final recipient can receive the full original key by XOR-combining the chain - applying the same bit-by-bit exclusive-or operation introduced in Appendix B, where a one and a zero combine to a one and matching bits cancel to a zero - but only if every relay in between is physically secure. If a spy can enter any one of those relay buildings, they can copy the key.

This is not a theoretical weakness. It is a real constraint that every currently deployed long-distance QKD network lives with. Governments and operators address it with physical security measures: guards, cameras, tamper sensors, access logs. Those measures are effective when properly implemented, but they replace quantum-physics security guarantees with conventional security practices. You are trusting the relay operator, the guards, and the building, not the laws of physics. The whole point of QKD is to back your security with physics, not promises.

The trusted-node problem also limits who can profitably deploy QKD. Building a chain of relay stations across a country requires control of every physical site along the chain. For a government operating on its own territory, that is difficult but manageable. For two private companies wanting to share keys across national borders, or for a bank trying to connect offices in cities served by different infrastructure providers, it becomes a serious obstacle. Quantum repeaters would change this picture entirely.

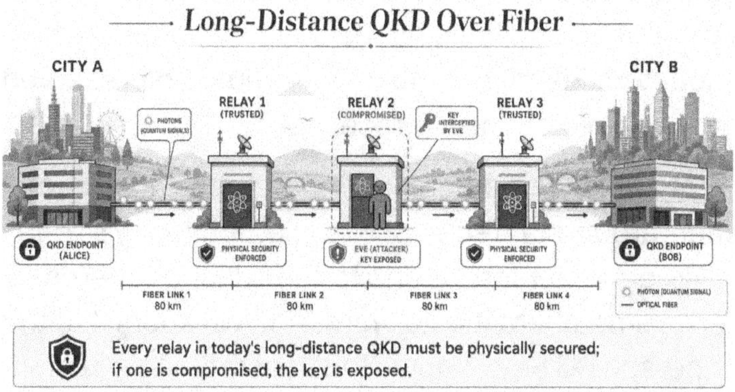

Diagram 12.1: The trusted-node chain and its vulnerability

The lesson is not that today's networks are broken. They are genuinely useful and genuinely more secure than ordinary digital relay networks for many threat models. The lesson is that trusted nodes represent a limitation, and the field knows exactly what it would take to remove it. Understanding that gap is what makes quantum repeaters so significant.

What a Quantum Repeater Would Do

The simplest way to understand a quantum repeater is to compare it with what today's relay stations do. A relay station today is a stop-and-convert station. The quantum signal comes in, gets measured (which collapses the quantum state into ordinary bits), and a new quantum signal goes out carrying those bits. The key exists briefly as a readable digital number inside that station. A quantum repeater would not do this. It would pass quantum information from one link to the next without ever collapsing the state into ordinary bits. The quantum identity of the signal would be preserved all the way from sender to receiver.

Think of it this way. Imagine you are passing a secret whisper around a table of people. In today's relay network, each person listens to the whisper, learns its content, remembers it as a word, and whispers it to the next person. They know the secret. A quantum repeater is more like a very unusual game of telephone, one in which the whisper travels around the table as a physical wave, never converted into a word in anyone's mind. Nobody in the chain learns the content. The only people who ever know it are the two people at the ends of the table.

Achieving this requires two technical capabilities working together. The first is quantum memory: the ability to receive a quantum state, hold it for a short time, and re-emit it faithfully without measuring it. The second is entanglement swapping: a procedure that uses quantum entanglement to connect two separate links into a single longer link without measuring the key information. Neither capability is trivial. Both require precise control of individual quantum systems, isolation from environmental noise, and hardware that is still being refined in labs around the world.

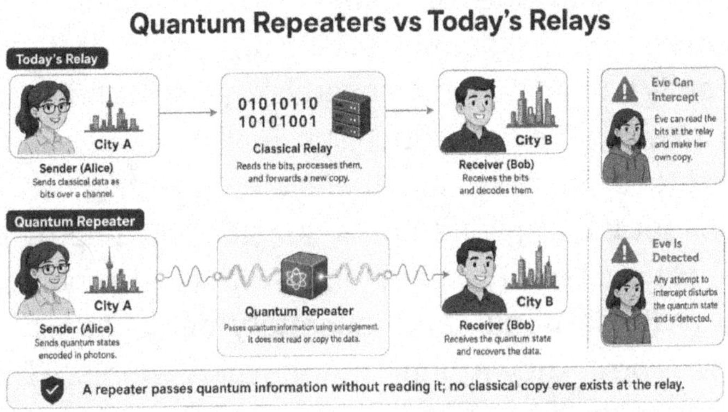

Diagram 12.2: Relay vs. repeater: two ways to extend a quantum link

When entanglement swapping is applied across a chain of quantum memories, something remarkable happens. Two photons that were never directly connected, one at the far-left end of the network and one at the far-right end, become entangled. The measurement occurs at the midpoints, but it measures a property that reveals nothing about the key. The key-bearing correlation is established at a distance by the physics of entanglement, not by any relay that knows the secret. This is the core mechanism that makes quantum repeaters a genuine solution to the trusted-node problem rather than just a better version of today's relay approach.

It is worth being honest: building a practical quantum repeater requires several technical components that are still maturing simultaneously. The quantum memory must work reliably, hold the state long enough to wait for the other link to become ready, and be compatible with the wavelengths that travel well through fiber. Entanglement swapping must succeed at sufficiently high rates for the overall key generation rate to be useful. The individual success probabilities of these steps multiply together, so improving each component by a small amount substantially improves the whole system. Researchers are making progress on all of these fronts, and laboratory demonstrations have already shown that the physics works. The challenge now is engineering.

Quantum Memory and Entanglement Links

Quantum memory is one of those concepts that sounds simple until you try to build it. What you want is a material or system that can absorb the quantum state of a photon, preserve that state undisturbed for a useful amount of time, and then release a photon carrying exactly the original state on demand. The difficulty is that quantum states are extraordinarily fragile. Any interaction with the environment, even a tiny vibration or a stray electromagnetic field, can cause decoherence, the technical word for the quantum state dissolving into a blur of randomness.

Several physical approaches are under active investigation. Some researchers trap individual atoms in laser beams and use the atoms' internal quantum states as memory. Others use rare-earth crystals, materials whose electrons can hold a quantum imprint at very low temperatures. Still others explore nitrogen-vacancy centers in diamond, tiny defects in the crystal structure where a single electron can be controlled with light. Each approach has trade-offs among storage time, efficiency, operating temperature, and compatibility with the wavelengths used by fiber networks. No single approach has yet won the race, which is itself a healthy sign: it means the field has multiple paths forward.

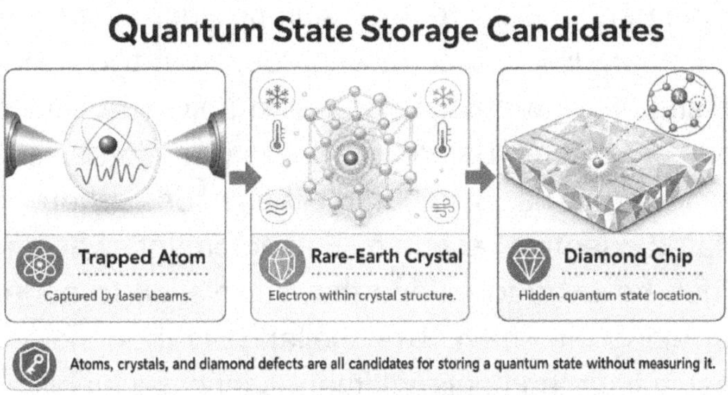

Diagram 12.3: Three approaches to quantum memory

The coherence time is the key performance number for quantum memory. Coherence time is the time during which a quantum system can maintain a quantum state before decoherence sets in and the information is lost. Think of it as the device's attention span: how long the memory can keep its mind on a single quantum state before that state slips away. Early demonstrations achieved coherence times of microseconds. More recent work has pushed into the millisecond range, and some specialized experiments have crossed the one-second threshold under highly controlled conditions. For a quantum repeater to work across, say, a 1,000-kilometer network, the memory must hold its state long enough for the entanglement signals from both sides to arrive and the swapping procedure to be completed. The longer the network, the more demanding the memory requirement. Current coherence times are sufficient for short-range demonstrations and are pushing toward the ranges needed for continental-scale links.

Entanglement links are the second ingredient. An entanglement link is a connection between two nodes that distributes an entangled photon pair, one photon going to each node, before any key information is sent. Once the pair is distributed, the nodes share a correlated quantum resource. Entanglement swapping can then be used to extend that correlation to nodes further along the chain. The challenge is that photons are lost in fiber, and entanglement cannot be amplified the way a classical signal can. This is where quantum error correction, still a very active research area, plays an important role. Short entanglement links with high success rates can be stitched together into longer paths by combining entanglement swapping with quantum error correction to detect and correct errors before they destroy the key.

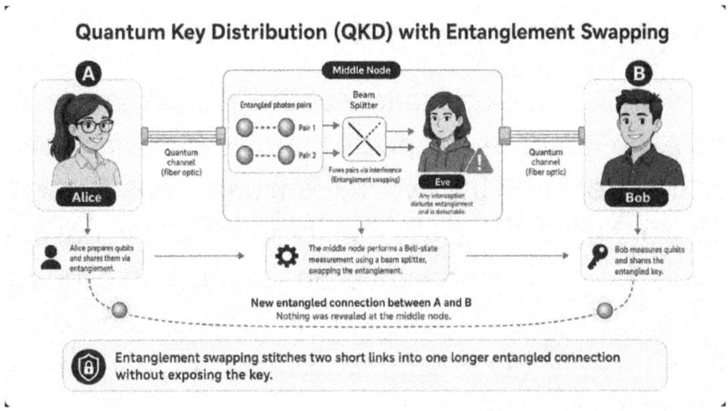

Diagram 12.4: Entanglement swapping across three nodes

The combination of quantum memory and entanglement swapping is sometimes described in the research community as a quantum repeater protocol. There are several generations of such protocols, differing mainly in how aggressively they use quantum error correction. The first generation uses only entanglement swapping with classical error detection. The second generation adds quantum error correction at the repeater nodes. The third generation uses full quantum error correction throughout, which would allow very long chains with high key rates but requires the most advanced hardware. Current laboratory demonstrations are primarily first-generation efforts, with second-generation experiments beginning to appear. The roadmap is clear, even if the timeline is not.

Entanglement-Based QKD

Most of this book has described QKD protocols in which Alice prepares a photon in a quantum state, sends it to Bob, and the two of them use measurement results to build a key. These prepare-and-measure protocols are the ones deployed in real networks today: BB84 and its variants, described in the early chapters. A different family of QKD protocols uses entanglement from the very beginning.

In an entanglement-based protocol, a source generates pairs of entangled photons. One photon from each pair goes to Alice, and the other goes to Bob. Neither Alice nor Bob has any control over the photons before they measure them. When they each make a random choice of measurement setting and record the outcome, the correlations between their outcomes, arising from entanglement, allow them to extract a shared key. The security argument is fundamentally different: it does not depend on what Alice's device does, because Alice did not prepare the photons. It depends on the correlations, which are certified by the laws of physics.

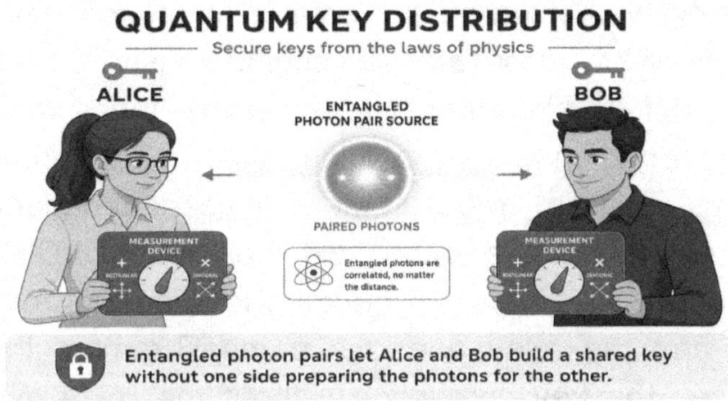

Diagram 12.5: Entanglement-based QKD: one source, two receivers

One important advantage of entanglement-based QKD is the possibility of device-independent security. You will recall from Chapter 10 that one of the trickier attack categories involves the quantum hardware itself: a device that behaves honestly under testing but leaks information when the key is being generated. Device-independent QKD is a protocol in which the security proof does not require you to trust the inner workings of Alice's detector or Bob's detector. Instead, the security is certified by the statistical test of the correlations between their measurements. If the correlation pattern satisfies a specific mathematical inequality (related to what physicists call a Bell inequality), then the physics of entanglement guarantees that no eavesdropper could have information about the key, regardless of what the devices are doing internally.

Device-independent QKD is not yet practical for real networks. It requires detectors with very high efficiency, very high-quality entangled photon pairs, and a very low error rate, all simultaneously. Current hardware falls short on at least two of those three requirements in most configurations. But partial versions, called measurement-device-independent (MDI) QKD, are already in field trials. In MDI-QKD, Alice and Bob each send photons to a central relay point, and the relay performs a measurement that produces the key correlation without the relay ever knowing the key. The relay can be completely untrusted. This removes one of the most dangerous attack surfaces in real deployments, the detector at Bob's site, from the security equation. Several metropolitan networks have demonstrated MDI-QKD links, and the technique is gradually moving from research to practice.

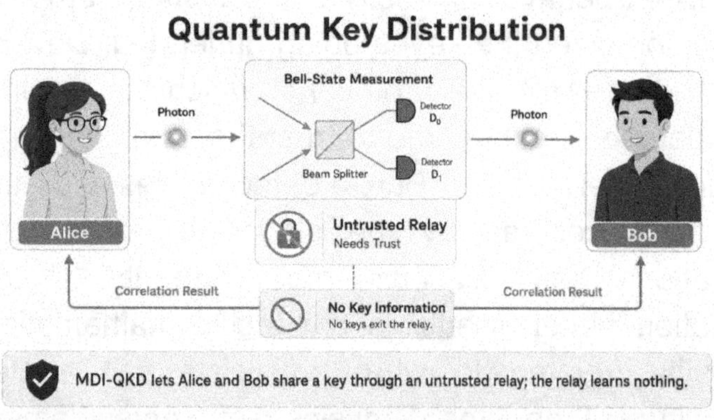

Diagram 12.6: Measurement-device-independent QKD

The broader family of entanglement-based protocols also includes twin-field QKD. This more recent protocol uses interference from photons on both sides to push the distance limit significantly beyond what standard

BB84 can achieve over fiber. Demonstrations of twin-field QKD have shown secure key generation over distances approaching 500 kilometers in a single fiber span without a relay. That is roughly three times the practical range of standard BB84 under similar conditions. Twin-field QKD is an active area of field trials and is likely to appear in real deployments within the next several years. It extends what is possible today without waiting for quantum repeaters to be perfected.

Entanglement-based protocols require more sophisticated hardware than prepare-and-measure protocols. Generating high-quality entangled photon pairs, maintaining their entanglement as they travel, and detecting them with high efficiency are all demanding tasks. But the security advantages they offer, especially in reducing the trust requirements on hardware, make them an important part of the future landscape. As hardware improves and costs come down, the line between what is deployed today and what is reserved for future entanglement-based systems will gradually shift.

Toward a Quantum Internet

The phrase "quantum internet" appears frequently in scientific papers, government strategy documents, and press releases. It is worth understanding precisely what it means, because it is more than just a faster or more secure version of the internet you use today.

A quantum internet would be a network that can transmit quantum information, not just classical bits, between distant nodes. Quantum information is the general case of the quantum states used in QKD, not just for key generation. A quantum internet could, in principle, distribute entanglement between any two points in the network on demand. It could support quantum computing tasks spread across multiple machines, quantum sensing experiments synchronized over long distances, and quantum communication protocols that go far beyond key sharing. It is, genuinely, a different kind of information infrastructure.

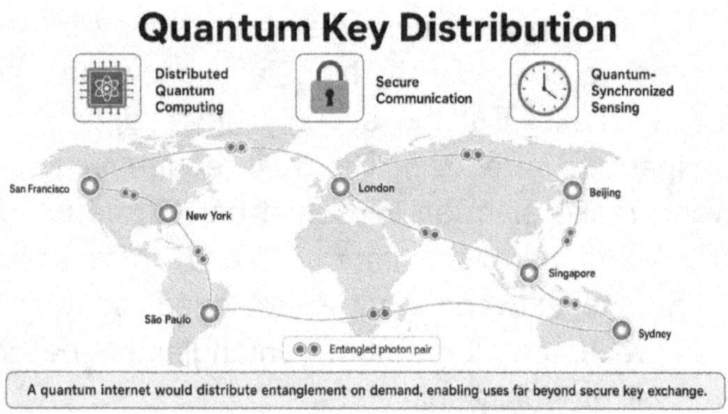

Diagram 12.7: The quantum internet vision

Several national programs have set out explicit roadmaps for building quantum networks that progressively add capability. The first stage is trusted-node QKD networks, which already exist. The second stage is entanglement-distribution networks, in which pairs of entangled photons can be delivered over metropolitan distances. The third stage is quantum repeater networks, in which entanglement can be extended over long distances without trusted nodes. The fourth stage is full quantum internet infrastructure, capable of supporting arbitrary quantum applications. Each stage requires the hardware advances of the previous stage. The world is currently between stages one and two in the most advanced national programs.

One of the most compelling near-term applications of a partial quantum internet is distributed quantum computing. Today's quantum computers are small and noisy. Connecting several of them with entanglement links could enable them to work together as a single, more capable system, much like classical computing clusters, where many machines share a computational task. This does not require a full planetary quantum internet. Even a metropolitan-scale quantum network connecting a handful of quantum processors could produce useful capabilities. Several research groups are actively pursuing exactly this goal, and some early demonstrations have connected two or three quantum processors over short distances using entanglement.

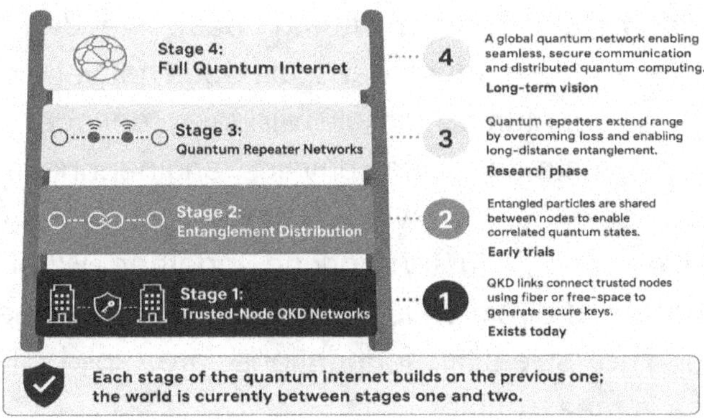

Diagram 12.8: Stages of quantum network development

Quantum sensing is another application that benefits from network-distributed entanglement. Atomic clocks synchronized via entanglement can, in principle, be more precise than any classical synchronization method. Telescopes spread across a continent can combine their observations via entanglement to achieve a resolution equivalent to that of a telescope the size of the continent itself. These are not science-fiction proposals; they are extensions of techniques already used in classical very-long-baseline interferometry, pushed into a new regime of precision by quantum correlations.

It would be dishonest to pretend that a global quantum internet is coming soon. The engineering challenges across all four stages are substantial. Each photon lost in fiber is a lost piece of quantum information that cannot be classically amplified to compensate. Each node must handle quantum states with extraordinary precision and low error rates. The materials science, engineering, and manufacturing base needed to produce quantum hardware at scale does not yet exist. Building a quantum internet will take decades of sustained investment across physics, engineering, and computer science, involving national governments, universities, and industry working together. What is clear is that the direction is correct and the investment is real. The scientific foundations are solid. The progress is measurable.

What Is Real, What Is Near, What Is a Vision

One of the most useful things this book can give you is a way to calibrate quantum claims. The field moves quickly, the stakes are high, and the gap between a laboratory demonstration and a commercial product can be enormous. Here is an honest picture of where things stand.

What is real today: QKD networks exist and are operating in several countries. Metropolitan fiber networks that use BB84-based protocols provide genuine security for government and financial communications across multiple cities. Satellite links have demonstrated secure key exchange over thousands of kilometers, though the ground stations must still be trusted nodes. Quantum random number generators are commercially available. Several national standards bodies have issued or are issuing standards for QKD equipment and protocols. The basic physics is fully understood and has been verified exhaustively.

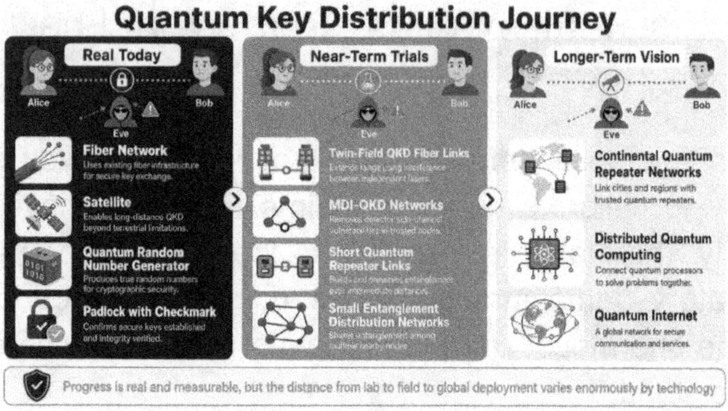

Diagram 12.9: The quantum security landscape today

What is near-term and in active trials: Twin-field QKD over longer fiber spans is moving toward deployment and will likely appear in real networks within a few years. MDI-QKD, which removes the untrusted-detector vulnerability, is in field trials and advancing. Short quantum repeater demonstrations over tens of kilometers are ongoing, with researchers reporting steady improvements in coherence times and entanglement success rates. Small entanglement distribution networks connecting two or three nodes are being built in university and government lab settings. Post-quantum cryptography, the classical software alternative discussed at length in Chapter 11, is being actively standardized and deployed, providing a practical near-term complement to QKD for most organizations.

What is a longer-term vision: Practical quantum repeater networks spanning hundreds of kilometers require hardware not yet available at production scale. Continental quantum networks without trusted nodes require repeaters to operate reliably across chains of many nodes, placing stringent performance requirements on each component. A quantum internet capable of distributing arbitrary quantum applications globally requires all of the above plus quantum processor nodes that do not yet exist outside small research settings. The timeline for these developments is measured in decades, not years.

A pattern worth watching: in the history of quantum information science, experiments once called impossible first became laboratory demonstrations, then engineering problems, and eventually products. The first single-photon detectors were finicky research tools. Today they are catalog items. The first quantum key exchange happened over a few centimeters of fiber. Today QKD links span hundreds of kilometers in the field. Quantum repeaters are following the same curve; the question is the speed, not the direction.

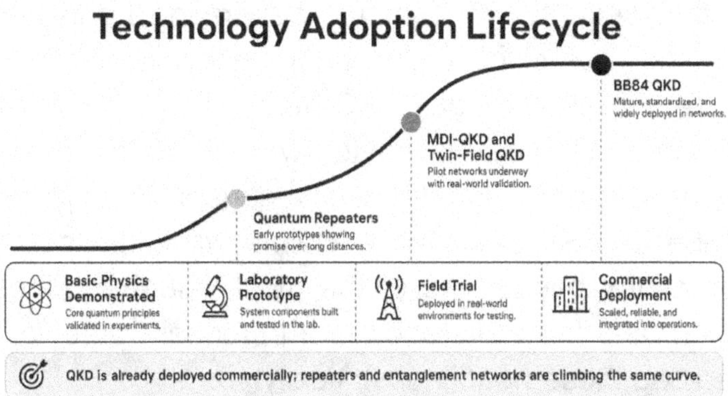

Diagram 12.10: How quantum technology matures: from lab to field

For managers and decision-makers reading this book, the practical implication is clear. Post-quantum cryptography is available now and should be the primary tool for most organizations updating their security infrastructure today. QKD is a valuable complement for high-sensitivity links and is deployable today in metropolitan settings. Quantum repeaters and the broader quantum internet are worth monitoring and understanding, but they are not yet inputs to a technology procurement decision. Tracking the field, understanding the vocabulary, and building relationships with researchers and vendors who specialize in this space will serve you far better than waiting for the perfect quantum solution before acting on any of it.

How to Keep Learning and Stay Grounded

Quantum security will continue generating headlines for years. Some of those headlines will describe genuine breakthroughs. Others will describe incremental laboratory results written up in press releases designed to attract investment. A few will describe entirely speculative claims with little or no experimental basis. One of the most valuable things you can take from this book is a personal framework for reading those headlines critically.

The first question to ask is: what exactly was demonstrated, and under what conditions? A quantum memory demonstration that works at four degrees above absolute zero in a vibration-isolated laboratory is real science. It is not a deployed product. A QKD system that achieves a record distance over fiber is a genuine milestone. It does not mean that the system is ready for your organization's infrastructure. The gap between a demonstration and a deployment is always worth probing.

The second question is: who conducted the work, and has it been independently replicated? Academic papers describing quantum experiments typically go through peer review, which is an imperfect but valuable filter. Results published in major physics and engineering journals have survived scrutiny from experts in the field. Press releases, vendor white papers, and government announcements have varying levels of technical rigor. Requiring that a claimed result appears in reviewed literature before treating it as established is a reasonable rule of thumb.

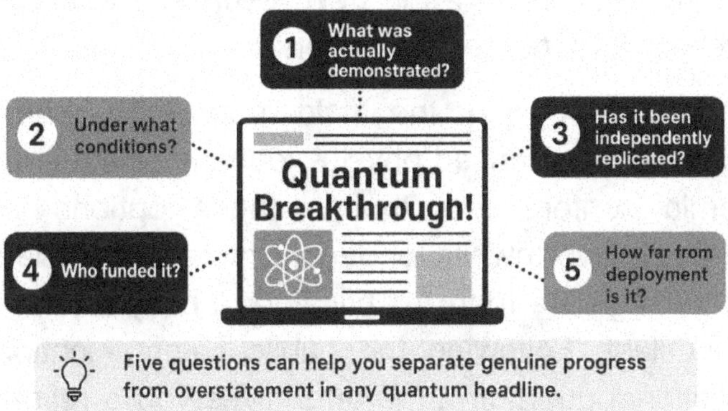

Diagram 12.11: Questions to ask when reading a quantum headline

The third question is: what is the gap between this result and a practical deployed system? A 500-kilometer fiber QKD demonstration is impressive and scientifically significant. A 500-kilometer demonstration with fiber costing $10,000 per kilometer and a detector requiring liquid helium cooling does not immediately translate into a product for enterprise customers. The cost, the size, the operating conditions, and the manufacturing scale all matter enormously in the translation from lab to deployment.

The fourth question is: what problem does this actually solve? New quantum results often come with ambitious claims about use cases. Tracing the logic from the result to the claimed application helps you evaluate whether the leap is reasonable. A new quantum random number generator does not, by itself, make existing cryptography stronger. A new entanglement demonstration does not automatically solve the trusted-node problem if the new result was achieved over ten meters, not ten kilometers.

Staying current with the field does not require subscribing to advanced physics journals or attending academic conferences, though those options are available and worthwhile for deeply interested readers. Several things are more practical for a busy manager or enthusiast. Following the public communications from national quantum programs gives you a reliable view of what governments are funding and prioritizing, which tends to be closer to practical deployment than academic cutting-edge work. Reading the technology coverage in trusted science journalism outlets, which translate peer-reviewed results into plain language, provides a useful filter. And engaging with communities of practitioners, security professionals who work with QKD systems and post-quantum algorithms in real organizations, gives you the ground-truth perspective that neither academic papers nor press releases provide.

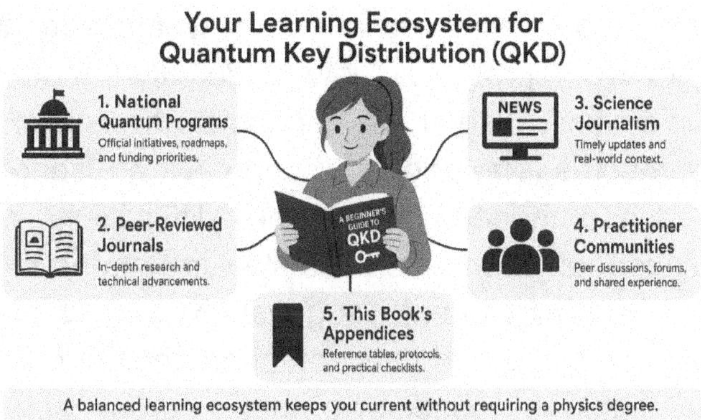

Diagram 12.12: Your learning ecosystem for quantum security

This book's appendices are your immediate next step. Appendix B walks through the BB84 protocol step by step, filling in any mechanistic detail you want to explore after reading the conceptual explanations in the earlier chapters. Appendix C lists free simulators and video resources where you can see QKD concepts demonstrated interactively, including several that require no software installation and run in a web browser. Appendix D provides a reading path organized by interest: if you want to go deeper into post-quantum mathematics, follow that path; if you want to explore quantum hardware and the physics of photons more carefully, follow a different one. The appendices are designed to be used in any order, according to what you want to explore next.

The field of quantum security is one of the most scientifically rigorous and practically consequential areas of technology development happening anywhere in the world today. It brings together physics that was considered purely theoretical twenty years ago and puts it to work protecting information that matters: communications, identities, financial transactions, medical records. The people working in it are serious, capable, and honest about the difficulties. The progress is real, even when it is slower than the most optimistic announcements suggest. You now have enough vocabulary and conceptual grounding to be a genuinely informed observer of that progress, able to ask the right questions, engage with the right sources, and make sensible decisions when quantum security issues arise within your organization.

That is a substantial takeaway from a book that started with the question of what makes a secret hard to steal. The path from that simple question to quantum repeaters, entanglement-based protocols, and a vision of a quantum internet has covered a lot of ground. You have traveled it without equations, without advanced physics prerequisites, and without losing the thread of what it all means for real people trying to protect real information in a world where the rules of cryptography are being rewritten. The road ahead is genuinely exciting. You are now ready to walk it.

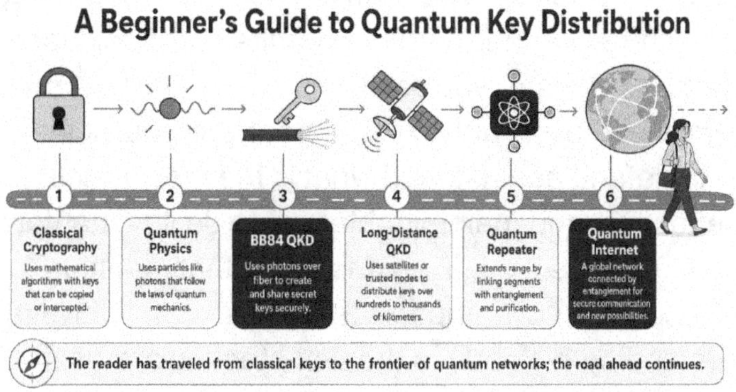

Diagram 12.13: The journey from classical keys to quantum networks

Post-quantum cryptography and QKD are not competing technologies; they are complementary tools. Chapter 11 explained this in detail. The honest summary for a decision-maker is simple. Post-quantum cryptography updates the mathematical locks that classical computers use, and it is now available in software at low cost and at scale. QKD uses the physics of light to distribute keys in a way that no computer, quantum or classical, can silently intercept. Both are worth deploying where appropriate, and neither is a reason to delay action on the other. Waiting for the perfect quantum solution is itself a risk, because the harvest-now-decrypt-later threat described in the earlier chapters does not wait for your procurement schedule.

The most grounded people in the quantum security field tend to share a few characteristics. They are excited by the science. They are honest about timelines. They are specific about what each technology actually does and does not do. They distinguish between laboratory results and deployable products. And they keep returning to the question that matters most in security: what threat are you actually trying to defend against, and what is the most practical, cost-effective, and maintainable way to do it? Quantum technology, at its best, is an answer to that question for a specific and important class of threats. Understanding it clearly is the first step toward using it well.

Key Concepts Checklist

• I understand why quantum repeaters would solve the trusted-node problem by relaying quantum information without measuring it

• I can describe, by analogy, how entanglement swapping extends a quantum link across multiple nodes

• I know that quantum memory and entanglement links are the two key components that repeaters require, and that both are still maturing in laboratories

• I understand what entanglement-based QKD adds, including the move toward device-independent security and the practical advance represented by MDI-QKD

• I can describe the quantum internet vision and its four developmental stages, and I know the world is currently between stages one and two

• I have a set of questions to ask when evaluating a quantum-security claim or headline

• I know where to go next using the appendices and the learning resources described in this chapter

Questions to Explore

• If quantum repeaters are eventually perfected, how would the network architecture described in Chapters 7 and 8 need to change, and what new security questions might arise at the repeater nodes themselves?

• Why is "stores and re-emits a quantum state without measuring it" the precise trick that breaks the trusted-node problem, and what would happen to the security guarantee if the memory made even one measurement during the relay process?

• A quantum internet would enable applications beyond secure key sharing. Which of those applications, distributed quantum computing, enhanced sensing, or something else, do you think would have the largest practical impact first, and why?

• When you next encounter a dramatic quantum-security headline, what will be the first question you ask, and how will you find a reliable source to check the claim against?

Try This / Reflect

Return to the personal "harvest now" inventory you created after reading the early chapters of this book. For each item on that list, whether it was a medical record, a financial account, a private message archive, or a professional communication, write a short paragraph describing what you now understand about its risk profile. Does the harvest-now-decrypt-later threat apply to it? Would QKD, post-quantum cryptography, or a combination of the two best address the risk? Is the threat urgent enough to justify action now, or is it something to monitor over the next few years? This reflection turns the book's entire conceptual journey into a practical planning exercise and gives you a concrete, personalized view of what quantum security means for your own world. Share the reflection with a colleague who has also read the book, compare your inventories, and see which risks and responses you agree on and which you evaluate differently. The conversation itself is part of the learning.

Glossary Terms

Quantum repeater - A device designed to extend QKD over long distances by relaying quantum information through entanglement swapping and quantum memory, without ever creating a readable classical copy of the key at the relay point.

Quantum memory - A physical system, such as a trapped atom, a rare-earth crystal, or a diamond defect, that can absorb the quantum state of a photon, hold it without measuring it, and re-emit it faithfully on demand.

Entanglement swapping - A quantum operation that combines two short entanglement links into one longer link by measuring the middle node that reveals no key information but transfers the entanglement correlation to the endpoints.

Entanglement-based QKD - A family of QKD protocols in which a shared source generates entangled photon pairs and distributes them to Alice and Bob, allowing them to extract a shared key from the correlations between their independent measurements.

Measurement-device-independent QKD (MDI-QKD) - A QKD protocol in which both Alice and Bob send photons to an untrusted central relay, which performs a measurement that produces the key correlation without learning any key information, removing the detector side-channel vulnerability.

Quantum internet - A future global network infrastructure capable of distributing quantum information, including entanglement, between distant nodes on demand, enabling applications beyond key exchange such as distributed quantum computing and enhanced quantum sensing.

Key Takeaways

• Quantum repeaters aim to eliminate the trusted-node weakness by relaying quantum information through entanglement swapping, never creating a readable key at any relay point

• They depend on two enabling technologies, quantum memory and entanglement swapping, both of which are progressing in laboratories but not yet ready for large-scale deployment

• Entanglement-based QKD protocols, including MDI-QKD and twin-field QKD, are already in field trials and extend the practical capabilities of today's networks by reducing hardware trust requirements and pushing distance limits

• The quantum internet is a genuine and well-funded long-term vision, currently between its first and second developmental stages, with a realistic but decades-long timeline to full realization

• You are now equipped to read quantum-security news critically, ask the right questions about any claimed advance, and use this book's appendices and recommended resources to continue learning on your own terms

13 Glossary

This glossary exists for one purpose: to give you a trustworthy place to land whenever a word stops you cold. Reading about quantum key distribution (QKD) means running into terms borrowed from physics, computer science, and network engineering, often all in the same paragraph. That overlap can make even a careful reader feel like the ground is shifting. The definitions here cut through that feeling by putting plain English first, every single time. No equations appear on these pages. Where a technical term has a useful everyday twin, the everyday version comes first.

The entries are grouped into five categories: Core Security Terms, Core Quantum Terms, Protocol Terms, Network and Hardware Terms, and Threat and Strategy Terms. Within each category, the terms appear in alphabetical order. That structure mirrors the way the chapters of this book build on each other: you start with the security concepts that motivate the whole project, move into the quantum physics that makes QKD possible, then work through the protocols and hardware, and finally face the threats that the whole field exists to defeat. You can read straight through or jump directly to any term you need.

A short note on cross-references. Many terms in one category refer to terms in another. Rather than cluttering every definition with arrows and pointers, the definitions use the related terms naturally, in context. If a word in a definition is itself unfamiliar, you can find it by scanning the category headers or using the alphabetical order within each section. Think of this glossary not as a wall of isolated facts but as a connected map of ideas, one that rewards both a quick lookup and a slow read.

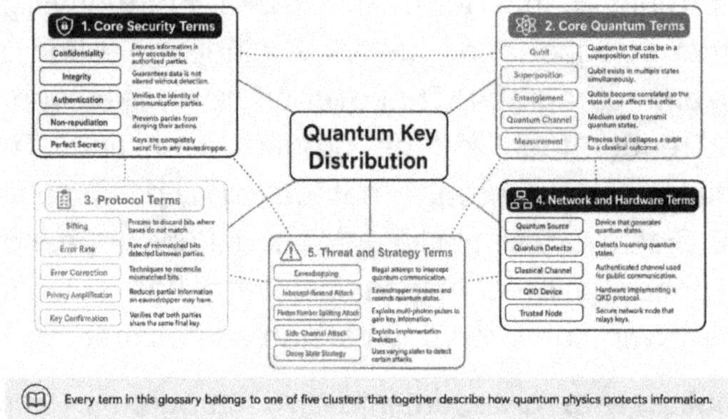

Diagram G.1: Map of quantum security vocabulary

Core Security Terms

Asymmetric encryption

Asymmetric encryption uses two mathematically linked keys instead of one: a public key that anyone may see, and a private key that only the owner keeps secret. Think of the public key as a padlock that anyone can snap shut on a box. Only the owner, who holds the matching private key, can open the box again. This design solves a puzzle that symmetric encryption cannot: how do you share a secret with someone you have never met? Asymmetric systems let two strangers exchange a locked box without ever exchanging the key itself. The most widely used asymmetric systems rely on mathematical problems that are easy to set up but very hard to reverse, such as multiplying two large prime numbers versus figuring out which primes were multiplied to produce a given result. QKD does not entirely replace asymmetric encryption but works alongside it, especially for authenticating the classical channel.

Authentication

Authentication is the act of proving that a message, file, or person is genuinely what it claims to be. It answers the question "Is this really from you?" rather than "Can anyone else read this?" A wax seal on an envelope was once an authentication device: breaking the seal indicated tampering. In digital systems, authentication often relies on a short piece of data called a message authentication code (MAC) or on a digital signature. In a QKD session, Alice and Bob must authenticate each other before they trust the key material they generate together. Without authentication, an attacker named Eve could pretend to be Bob to Alice and Alice to Bob, sitting in the middle of the conversation unseen.

Cipher

A cipher is a specific method or algorithm for scrambling and unscrambling information. The word comes from the Arabic word for zero, sifr, reflecting the long history of numerical methods in secret writing. A Caesar cipher shifts every letter of the alphabet by a fixed number of positions; a modern cipher like AES (Advanced Encryption Standard) uses mathematical operations across blocks of bits. The cipher is the recipe; the key is the secret ingredient that makes each batch of ciphertext unique. The same cipher can produce completely different ciphertext from the same plaintext if the key changes.

Ciphertext

Ciphertext is the scrambled output produced when a cipher processes readable information. Before processing, the information is called plaintext. After processing, it looks like a random string of characters or bits. The word "ciphertext" is simply the technical name for "the locked version." Decrypting the ciphertext with the correct key returns the original plaintext. Without the key, a strong cipher turns ciphertext into noise that is effectively impossible to reverse.

Encryption

Encryption is the process of converting readable information (plaintext) into an unreadable form (ciphertext) so that only people holding the correct key can reverse it. Think of it as locking a document inside a box with a combination lock. The document is still there; you cannot read it without the combination. Encryption is the core tool that protects data in transit (crossing a network) and data at rest (sitting on a hard drive). QKD provides one ingredient that encryption requires: a shared secret key. Once Alice and Bob hold a matching key, they use it with a conventional cipher to encrypt their actual messages.

Hash function

A hash function is a mathematical operation that takes any piece of data and produces a fixed-length "fingerprint" from it. Change even one letter of the input and the fingerprint changes completely. Hash functions are one-way: you can easily go from data to a fingerprint, but you cannot reconstruct the original data from the fingerprint alone. In QKD, hash functions appear in privacy amplification, the step in which Alice and Bob compress their raw key bits to eliminate any partial information Eve may have gained from observing the quantum channel. They also appear in message authentication codes.

Key

A key is a secret piece of information that controls how a cipher locks or unlocks data. In physical life, a key is a shaped piece of metal that fits one specific lock. In digital cryptography, a key is a string of bits, often dozens or hundreds of characters long. The security of most encrypted communication rests entirely on the secrecy of the key, not on the secrecy of the cipher algorithm itself. This is a deliberate design principle: if the algorithm were the secret, discovering it would break all messages ever encrypted with it. Keeping only the key secret means you can publish the algorithm to the world, let experts try to attack it, and still keep your messages safe as long as the key stays hidden. QKD is specifically a method for distributing keys, not for encrypting messages directly.

Key rate

Key rate is the speed at which a QKD system produces usable secret key bits, measured in bits per second or sometimes kilobits per second. A higher key rate means the system can encrypt more data or refresh the key more often. The key rate depends on many factors: the distance between Alice and Bob, the quality of their single-photon detectors, the amount of light lost in the fiber or free-space channel, and the channel's noise. Today's practical QKD systems typically achieve key rates ranging from a few kilobits per second over tens of kilometers to just a handful of bits per second over hundreds of kilometers.

One-time pad (OTP)

The one-time pad is the only encryption scheme mathematically proven to be perfectly secure, with no computational shortcuts possible. It works by combining each bit of a message with a corresponding bit from a random key, using a simple operation called XOR (exclusive or). The key must be exactly as long as the message and must never be reused. The catch is obvious: if you have a secure way to share a key for every message you ever send, why not just use that secure channel to send the message directly? QKD partially solves this problem by providing a way to distribute fresh random key bits at high volume. When Alice and Bob have a fast QKD link, they can continuously replenish their one-time pad supply and use it for the most sensitive communications.

Private key

A private key is the secret half of an asymmetric key pair. It never leaves its owner's possession. In RSA, for example, the private key is a large number derived from two secret prime numbers. Anyone who holds the private key can decrypt messages encrypted with the corresponding public key and sign messages to prove their identity. The security of asymmetric cryptography depends entirely on the infeasibility of computing the private key from the public key. Shor's algorithm, which a sufficiently powerful quantum computer could run, threatens to break that infeasibility for certain types of private keys.

Public key

A public key is the freely shareable half of an asymmetric key pair. You can post your public key on a website, include it in an email signature, or upload it to a key server. Anyone who wants to send you a private message locks it with your public key; only your private key can unlock it. Public keys are also used to verify digital signatures: a message signed with a private key can be verified by anyone who holds the corresponding public key, thereby confirming the sender's identity without revealing the private key.

Symmetric encryption

Symmetric encryption uses a single key for both locking and unlocking. Both the sender and the receiver must hold a copy of the same key. Think of a physical safe to which two people each have an identical key. The challenge is key distribution: how do you safely share the key with your intended partner without letting anyone else see it? This is precisely the problem QKD solves. Once Alice and Bob have agreed on a symmetric key via QKD, they can use a symmetric cipher such as AES to encrypt and decrypt messages quickly and efficiently. Symmetric ciphers are generally much faster than asymmetric ones, so symmetric encryption handles the bulk of real-world data; asymmetric cryptography handles the key exchange step.

Core Quantum Terms

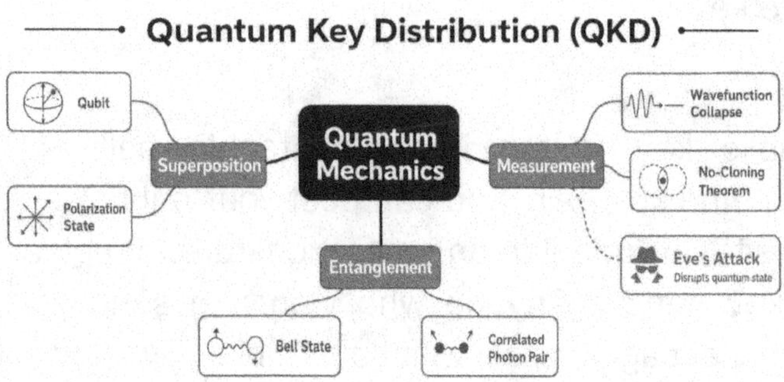

Six core quantum concepts and how they connect to each other in QKD.

Diagram G.2: Core quantum concepts vocabulary tree

Bell state

A Bell state is a specific kind of entangled state shared between two qubits (or two photons). The name honors physicist John Bell, who devised tests to prove that entangled particles really do share quantum correlations rather than carrying hidden instructions set at the moment of creation. There are four Bell states, each representing a different pattern of correlation between the two particles. In entanglement-based

QKD protocols, Alice and Bob share photon pairs in a Bell state and then measure each photon independently. Their results, though random individually, are correlated in a way that allows them to build a shared secret key, while any eavesdropper necessarily disturbs those correlations.

Entanglement

Entanglement is a quantum relationship between two or more particles in which their properties are correlated no matter how far apart they are. Measure one particle, and you instantly know something about its partner, even if the partner is on the other side of the planet. This is not a hidden radio signal; Einstein called it "spooky action at a distance" precisely because it cannot be explained by any conventional message traveling between the particles. Entanglement does not allow faster-than-light communication, because you cannot control which random outcome a measurement will produce. What it provides is correlated randomness, which is the foundation of certain advanced QKD protocols. An entangled photon pair is like a pair of magic coins: flip each one in separate rooms, and they always land opposite faces, yet each flip is individually random and unpredictable.

Measurement

In everyday life, measuring something does not change it. You can weigh a bag of coffee without altering the coffee. In quantum mechanics, measurement is fundamentally different: the act of measuring a quantum object forces it to "choose" a definite value, and this choice is irreversible. Before measurement, the object may exist in a superposition of multiple possible values. Measurement collapses that superposition into one specific outcome. In QKD, this principle is a security feature rather than a limitation. If Eve tries to read a photon's polarization to learn the key bit it carries, she must measure it. That measurement disturbs the photon, introducing detectable errors in Alice and Bob's results. The disturbance is the alarm signal.

No-cloning theorem

The no-cloning theorem is a mathematical proof that it is impossible to make a perfect copy of an unknown quantum state. You can copy a classical bit easily: read it (0 or 1) and write that value somewhere else. But a quantum bit (qubit) can exist in a superposition of states that you cannot read without disturbing it, and you cannot duplicate the full state without first reading it. The no-cloning theorem prevents Eve from intercepting photons, copying their quantum state, forwarding the originals to Bob, and reading her copies later at leisure. Any attempt to copy the quantum information necessarily disturbs it, leaving a fingerprint of the intrusion in the error statistics.

Photon

A photon is a single particle of light. It is the smallest possible unit of electromagnetic radiation. Photons travel at the speed of light and carry energy that depends on their frequency or color. In QKD, individual photons carry key bits. A single photon's polarization (the direction in which its electric field oscillates) encodes a 0 or a 1. Because photons are quantum particles, they obey quantum rules: measuring them can disturb them, and you cannot copy them without detection. Using single photons is what gives QKD its physical security guarantee. Classical optical communication uses pulses containing billions of photons; QKD works at the extreme limit of one photon per pulse (or a carefully calibrated small number in decoy-state protocols).

Polarization

Polarization describes the orientation of the electric-field oscillation of a photon. Imagine a rope tied to a fence. You can shake the rope up and down (vertical polarization), side to side (horizontal polarization), or diagonally. Polarized sunglasses block one orientation and pass the other, which is why they reduce glare from water or pavement. In QKD, specific polarization orientations represent key bit values. A photon polarized vertically might represent a 0; one polarized horizontally might represent a 1. Diagonal orientations represent bits in a second basis. Alice controls the polarization she sends; Bob tries to measure it; the quantum rules determine when their choices align and when they do not.

Qubit

A qubit (quantum bit) is the basic unit of quantum information, analogous to the classical bit in ordinary computing. A classical bit is always either 0 or 1. A qubit can be in a superposition of both 0 and 1 simultaneously until it is measured, at which point it collapses to one value with a certain probability. In QKD, each photon sent from Alice to Bob carries one qubit of information in its polarization or phase. The qubit's quantum properties (superposition, entanglement, and resistance to cloning) are what give QKD its security advantage over purely classical key exchange. A quantum computer manipulates many qubits at once to perform calculations that would take classical computers an impractically long time.

Superposition

Superposition is the quantum property that allows a particle to exist in a combination of multiple states at the same time until a measurement is made. A classical light switch is either on or off. A qubit in superposition is, in a precise mathematical sense, both on and off simultaneously, with a probability associated with each outcome. When you measure the qubit, you get one definite result (on or off) with that probability. The superposition collapses. This is not a trick of incomplete knowledge. The particle genuinely does not have a single definite value until the measurement forces it to acquire one. Superposition is why quantum computers can explore many computational paths at once, and it is why Eve cannot silently read a photon's state without disturbing it.

Wavefunction

A wavefunction is the mathematical description of a quantum particle's state. It encodes the probabilities of all possible measurement outcomes for the particle.

Before measurement, the wavefunction can be a superposition; after measurement, it collapses to reflect the definite outcome obtained. You do not need to work with the mathematics of wavefunctions to understand QKD, but you will occasionally see the term in more technical discussions. Think of a wavefunction as a probability cloud: it describes where a particle "might be" or "might be oriented" before you look—looking forces the cloud to condense into a single point.

Protocol Terms

Alice

Alice is the conventional name for the sender in a cryptographic or QKD protocol. She wants to share a secret key with Bob. The names Alice, Bob, and Eve are universally used in cryptography literature and make it easier to follow who is doing what in a protocol description. In this book, Alice typically operates the QKD transmitter: she generates random bit values, encodes them as photon states, and sends them to Bob.

Basis

In QKD, a basis is the set of measurement orientations (or encoding orientations) that a party uses for a particular photon. In the BB84 protocol, there are two bases: the rectilinear basis (vertical and horizontal polarizations) and the diagonal basis (two diagonal polarizations at 45 ° and 135°). Alice randomly chooses a basis for each photon she sends. Bob independently and randomly chooses a basis for each measurement.

When they happen to choose the same basis for a photon, their results should agree (allowing for noise). When they choose different bases, the result is random and must be discarded. The basis-matching step is called sifting.

BB84

BB84 is the first and most widely known QKD protocol. The name comes from its inventors, Charles Bennett and Gilles Brassard, and the year of its publication, 1984. The protocol uses four polarization states divided into two bases. Alice sends photons one by one with randomly chosen polarizations; Bob measures each photon in a randomly chosen basis; they compare their basis choices over a public classical channel; they keep only the results where their bases matched (sifting); they estimate the error rate to check for eavesdropping; and they use error correction and privacy amplification to convert their noisy matching bits into a shorter, clean, secret key. BB84 is proven secure against any attack allowed by quantum physics, provided the implementation follows the protocol exactly.

Bob

Bob is the conventional name for the receiver in a cryptographic or QKD protocol. He wants to receive the secret key that Alice is sending. In this book, Bob operates the QKD receiver: he measures photons as they arrive, records his basis choices and outcomes, and participates in the classical post-processing steps (sifting, error correction, privacy amplification) that turn raw measurement results into a usable key.

Decoy states

Decoy states are a practical technique that makes the BB84 protocol (and similar protocols) secure against photon-number-splitting attacks. Real laser sources occasionally emit pulses with more than one photon. An attacker could keep one of those extra photons and measure it later once Alice and Bob have revealed their basis choices. Decoy states defeat this attack by mixing in pulses of different average photon numbers at random. By comparing the statistics across different decoy intensities, Alice and Bob can detect whether any multi-photon pulses were intercepted. Decoy states are now standard in commercial QKD systems and are part of most current security proofs.

E91

E91 is an entanglement-based QKD protocol proposed by Artur Ekert in 1991. Instead of Alice sending photons with known polarizations, E91 uses a source that produces pairs of entangled photons, sending one photon to Alice and one to Bob. Alice and Bob each measure their photon independently and randomly. The correlations between their results, tested using Bell's theorem, confirm both that the photons were genuinely entangled and that no eavesdropper has interfered. E91 is more technically demanding than BB84 because it requires a reliable source of entangled pairs. Still, it offers a stronger, more direct security argument grounded in the fundamental physics of quantum correlations.

Error correction

Error correction is the post-processing step in which Alice and Bob reconcile the differences in their raw key bits caused by channel noise or eavesdropping. Even a perfect quantum channel has some natural noise; photons occasionally arrive with the wrong orientation due to environmental effects. Alice and Bob compare a subset of their bits over a public classical channel (which reveals some information) to estimate the error pattern and correct their copies to match. The bits used for comparison are discarded. The result is a shorter, matching string of bits called the reconciled key. Error correction is followed by privacy amplification.

Eve

Eve is the conventional name for the eavesdropper in a cryptographic or QKD protocol. Her name is a deliberate reference to "eavesdropping." In security analyses, Eve is assumed to be maximally capable: she controls the quantum channel entirely, can intercept and resend every photon, has unlimited computing power, and can store anything she cannot read immediately. By designing protocols that remain secure even against this all-powerful Eve, researchers ensure that real-world attackers (who are always weaker than the theoretical maximum) cannot break the protocol either.

Privacy amplification

Privacy amplification is the final step in QKD post-processing. After error correction, Alice and Bob share matching key bits, but Eve may have partial information about some of those bits from her observations of the quantum channel. Privacy amplification uses a hash function to compress the reconciled key into a shorter string from which Eve's partial information has been mathematically squeezed out. The amount of compression depends on estimates of how much information Eve could have gained. The result is a shorter but provably secret key. Privacy amplification turns a key that Eve "mostly cannot read" into a key that she "cannot read at all, with overwhelming probability."

QBER (quantum bit error rate)

The quantum bit error rate (QBER) is the fraction of received key bits that do not match what Alice sent, expressed as a percentage. A QBER of 3% means that 3 out of every 100 bits arrived with the wrong value. Some error is always expected from ordinary channel noise and imperfect detectors. A QBER above a certain threshold (typically around 11% for BB84) indicates that Eve may have been intercepting photons, because her intercept-and-resend strategy inevitably introduces errors. Alice and Bob continuously monitor the QBER; if it exceeds the threshold, they abort the session and try again later.

Sifting

Sifting is the step in BB84 and similar protocols in which Alice and Bob compare their basis choices over the public classical channel and discard every photon for which they used different bases. Because Alice and Bob each choose their basis randomly and independently, they will agree on roughly half of all photons (in the two-basis BB84 scheme). The bits from the other half contain no useful information and are discarded. Sifting typically halves the raw bit pool. The surviving bits form the sifted key, which then goes through error correction and privacy amplification to become the final secret key.

Network and Hardware Terms

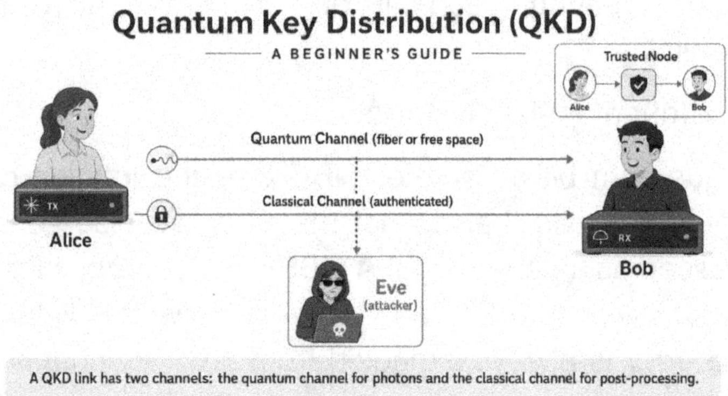

Diagram G.3: QKD network hardware vocabulary map

Calibration

Calibration is the process of precisely aligning the transmitter and receiver components of a QKD system so that Alice's polarization encodings match Bob's measurement settings. If Alice sends a vertically polarized photon but Bob's detector is tilted slightly off-axis, he will measure the wrong result more often than expected, raising the QBER artificially. Regular calibration is essential for keeping the QBER low and the key rate high. Calibration also involves verifying that the single-photon detectors respond correctly and that timing signals are synchronized. An attacker can sometimes exploit the calibration process itself; this is one category of implementation attack.

Classical channel

The classical channel is the ordinary communication link (such as a standard internet or fiber connection) that Alice and Bob use to exchange the non-secret parts of QKD post-processing, including basis comparisons during sifting, error-correction messages, and confirmation of privacy amplification parameters. The classical channel is called "classical" to distinguish it from the quantum channel. It is not secret; anyone, including Eve, can listen in. It must, however, be authenticated, meaning Alice and Bob must verify that the messages they receive on the classical channel genuinely come from each other and have not been altered. Without authentication of the classical channel, Eve could mount a man-in-the-middle attack.

Dark count

A dark count is a false detection event in a single-photon detector: the detector fires and registers a "photon arrived" signal even when no photon was actually present. Dark counts arise from thermal noise within the detector material. Every real single-photon detector has a dark count rate, measured in counts per second. Dark counts add noise to the QKD signal, increasing the effective QBER. Cooling the detector to very low temperatures significantly reduces dark counts, which is why many high-performance QKD systems use detectors chilled with liquid helium or specialized refrigerators.

Denial of service

A denial-of-service (DoS) attack does not try to steal the key; it tries to prevent Alice and Bob from generating a key at all. An attacker can flood the quantum channel with bright light, physically cut the fiber, or jam the classical channel to disrupt post-processing. QKD does not protect against DoS in the same way it protects against eavesdropping. Defense against DoS requires redundant paths, physical security of the fiber, rate-limiting of the classical channel, and monitoring for anomalous conditions. DoS attacks are a real practical concern for QKD network operators, distinct from the cryptographic threat of key compromise.

Fiber loss

Fiber loss is the gradual reduction in the strength of a light signal as it travels through an optical fiber, measured in decibels per kilometer. Standard telecommunication fiber loses about 0.2 decibels per kilometer. At the single-photon level used in QKD, this loss is critical. Every photon that is absorbed or scattered by the fiber is a key bit lost forever, and you cannot amplify the signal without violating quantum rules. Fiber loss is the primary factor that limits QKD to practical distances of roughly 100 to 200 kilometers without trusted nodes or quantum repeaters. Below about one photon per second arriving at Bob's detector, the key rate becomes too slow for most applications.

Free-space link

A free-space link is a QKD optical connection that transmits photons through air (or vacuum) rather than through an optical fiber. Free-space links are used for terrestrial line-of-sight connections between buildings and for satellite-based QKD, where photons travel through the atmosphere and the vacuum of space. The losses and noise sources in a free-space link differ from those in fiber: atmospheric turbulence, pointing and tracking errors, and daytime background light (sunlight) are the main challenges. Free-space links have achieved QKD over hundreds of kilometers using satellites in orbit and over tens of kilometers between mountain observatories.

Ground station

A ground station is the terrestrial endpoint of a satellite-based QKD link. It contains a large optical telescope to track the satellite and collect incoming photons, along with single-photon detection electronics and classical communication equipment for post-processing. Ground stations must compensate for atmospheric turbulence and Doppler shifts in real time as the satellite passes overhead at orbital speed. A typical low-Earth-orbit satellite is above the horizon for only a few minutes per pass, so the key generation window is short and efficient tracking is essential.

Implementation attack

An implementation attack exploits flaws in the physical hardware or software of a QKD system rather than flaws in the underlying quantum protocol. The protocol's security proof assumes ideal devices: perfectly random photon sources, perfectly efficient detectors, and so on. Real devices fall short of the ideal, and those gaps can be exploited. A blinding attack, for example, floods the single-photon detector with bright light to force it into a classical regime where Eve can intercept photons without triggering the usual disturbance statistics. Implementation attacks are a major area of current QKD research, and device-independent QKD is a response to them: it aims to provide security even when the devices are untrusted.

Key management system

A key management system (KMS) is the software infrastructure that stores, distributes, and tracks the secret keys produced by a QKD device or any other key source. Once Alice and Bob's QKD hardware has generated a batch of secret key bits, the KMS receives those bits, stores them securely, and delivers them to the applications that need encryption. A KMS also handles key rotation (replacing old keys with new ones), key expiration, logging who used which key, and emergency key revocation. QKD generates the raw material (random secret bits); the KMS turns that raw material into a managed cryptographic resource.

Quantum channel

The quantum channel is the physical path over which Alice sends photons to Bob. In fiber-based QKD, the quantum channel is an optical fiber. In free-space QKD, the line of sight is through air or vacuum. The quantum channel is called "quantum" because the signal it carries (single photons or entangled photon pairs) obeys quantum physics. Any eavesdropping on the quantum channel necessarily disturbs the photons it carries. The quantum channel is unidirectional in most protocols: Alice sends photons to Bob, not the other way around. It is separate from the classical channel over which post-processing messages travel.

Quantum memory

Quantum memory is a device that can store a quantum state (such as the polarization of a photon or the state of an entangled particle) for some period of time without destroying it. Quantum memories are a key enabling technology for quantum repeaters: a repeater station needs to hold one half of an entangled pair. At the same time, it waits for the other half to arrive from the other direction. Present-day quantum memories have storage times ranging from microseconds to seconds, depending on the physical platform. Making quantum memories practical for real-world networks is one of the major engineering challenges of the field.

Quantum repeater

A quantum repeater is a device designed to extend the range of QKD beyond the limits imposed by fiber loss and the no-cloning theorem. In classical networks, a signal that grows too weak can be amplified: copy the bits, boost the power, and send them onward. In quantum networks, you cannot copy an unknown quantum state (the no-cloning theorem forbids it). A quantum repeater instead uses entanglement swapping: it creates two short entangled links. Then it performs a measurement that "joins" them into one longer link, without ever copying the quantum state. Chains of quantum repeaters could, in principle, extend QKD across continents. Building practical quantum repeaters is one of the hardest open problems in quantum networking.

Satellite QKD

Satellite QKD uses an orbiting satellite to relay photons (or entangled photon pairs) between two ground stations that are too far apart for a direct fiber or free-space link. The satellite acts as a trusted node in a key relay protocol, or as a source of entangled pairs distributed simultaneously to two ground stations. Satellite QKD has been demonstrated over intercontinental distances, proving that the concept works in practice. The main limitations are weather (clouds block the optical beam), the short key-generation window per satellite pass, and the cost and complexity of deploying space-qualified optical hardware.

Single-photon detector

A single-photon detector is a highly sensitive electronic device capable of registering the arrival of a single photon. It must be sensitive enough to detect a single quantum particle of light while fast enough to time its arrival to within a fraction of a nanosecond. Common types include avalanche photodiodes (APDs) operated in Geiger mode and superconducting nanowire single-photon detectors (SNSPDs). SNSPDs achieve very high efficiency and very low dark counts but require cooling to temperatures near absolute zero. Single-photon detectors are among the most demanding components in a QKD system and are a major focus of hardware improvement.

Trusted node

A trusted node is a relay station in a QKD network that extends the system's reach beyond the range of a single quantum link. Alice and Node 1 share a QKD key; Node 1 and Bob share a different QKD key. Node 1 uses its two keys to relay the secret between Alice and Bob: it decrypts Alice's key material using the first shared key and re-encrypts it using the second. The security of this relay depends entirely on the node's physical and administrative security. If an attacker compromises the trusted node, the whole chain is broken. Trusted nodes are a practical workaround used today; quantum repeaters are the long-term alternative that would eliminate the need to trust intermediate nodes.

Threat and Strategy Terms

Code-based cryptography

Code-based cryptography is a family of post-quantum cryptographic algorithms whose security relies on the difficulty of decoding a random linear error-correcting code. Error-correcting codes are used in digital communications to detect and correct transmission errors; the hard mathematical problem here is essentially the reverse: given a garbled codeword, determine which errors were introduced. This problem has been studied since the 1970s, and no quantum algorithm is known to solve it significantly faster than classical algorithms. Code-based schemes tend to produce large key sizes, which is their main practical drawback, but they benefit from decades of cryptanalytic scrutiny.

Defense in depth

Defense in depth is a security strategy that layers multiple independent protective measures so that an attacker who defeats one layer still faces several others. The term comes from military doctrine: a defending army does not rely on a single line of fortifications but on multiple overlapping lines. In the context of quantum-safe security, defense-in-depth might mean running QKD alongside post-quantum cryptography, adding strong authentication, physically securing cables and server rooms, monitoring for anomalous network activity, and maintaining an incident response plan. No single technology or measure is perfect; the combination of layers is what makes the overall system resilient.

Grover's algorithm

Grover's algorithm is a quantum algorithm that searches an unsorted list of N items in roughly sqrt (N) steps, compared to N steps on a classical computer. In cryptography, the most important implication is its effect on symmetric ciphers and hash functions: a quantum computer running Grover's algorithm can search through all possible keys roughly twice as fast, effectively halving the key length. A 128-bit symmetric key becomes roughly as secure as a 64-bit key against a quantum-equipped attacker. The standard response is to double symmetric key lengths: 256-bit AES is considered quantum-safe against Grover's algorithm. Unlike Shor's algorithm (which breaks certain asymmetric cryptosystems entirely), Grover's algorithm poses a manageable threat that can be addressed with larger key sizes rather than by replacing the algorithm.

Harvest-now-decrypt-later

Harvest-now-decrypt-later (HNDL) is an attack strategy in which an adversary records large volumes of encrypted traffic today, stores the ciphertext, and waits until a sufficiently powerful quantum computer becomes available to decrypt it. The attack exploits the fact that much of today's encryption uses asymmetric algorithms that Shor's algorithm could eventually break. Data with a long secrecy lifetime (diplomatic communications, medical records, intellectual property) is already at risk under this strategy, even before a cryptographically relevant quantum computer exists. HNDL is the primary reason why the migration to quantum-safe cryptography is urgent today rather than a problem for a decade from now.

Hash-based cryptography

Hash-based cryptography is a family of post-quantum cryptographic algorithms (primarily digital signature schemes) whose security relies solely on the properties of hash functions. Hash functions are well-understood mathematical tools with no known quantum speedup beyond Grover's algorithm (which is manageable). Hash-based signatures like XMSS and SPHINCS+ have been studied for decades and carry a reputation for conservative, well-understood security. Their main limitation is that they produce relatively large signatures compared to classical schemes, and some variants have restrictions on how many times a single key pair can be used.

Lattice-based cryptography

Lattice-based cryptography is a family of post-quantum cryptographic algorithms whose security relies on the mathematical difficulty of finding short vectors in a high-dimensional geometric structure called a lattice. The problems are believed to be hard for both classical and quantum computers, with no known efficient quantum algorithm. Lattice-based schemes are currently the leading candidates in standardization processes for post-quantum cryptography because they offer a balance of reasonable key and signature sizes, fast computation, and strong theoretical security arguments. Several lattice-based algorithms have been selected for standardization by national standards bodies.

Man-in-the-middle attack

A man-in-the-middle (MITM) attack occurs when an attacker secretly positions herself between two communicating parties, intercepting and potentially altering all traffic. In contrast, each party believes she is communicating directly with the other. In QKD, the no-cloning theorem and disturbance-based detection protect against a quantum MITM on the quantum channel. However, if the classical channel is not properly authenticated, Eve can perform a MITM attack by pretending to be Bob to Alice and Alice to Bob on the classical channel, establishing two separate QKD sessions and learning both keys. This is why authentication of the classical channel is an absolute requirement in any QKD deployment.

Multivariate cryptography

Multivariate cryptography is a family of post-quantum cryptographic algorithms (mainly digital signature schemes) whose security relies on the difficulty of solving systems of multivariate polynomial equations over a finite field. Solving a large random system of such equations is believed to be computationally hard for both classical and quantum computers. Multivariate schemes tend to be fast to verify and produce small signatures, making them attractive for constrained environments. Their key generation can be large, and advances in algebraic cryptanalysis have broken some multivariate schemes, so careful selection of parameters is important.

Post-quantum cryptography (PQC)

Post-quantum cryptography (PQC) refers to classical cryptographic algorithms (running on ordinary computers) designed to resist attacks by quantum computers. The goal is to replace the asymmetric algorithms (RSA, elliptic-curve cryptography) that Shor's algorithm would break with new algorithms based on mathematical problems that are hard even for quantum computers. PQC is distinct from QKD: QKD uses quantum physics to distribute keys, while PQC uses clever mathematics to make conventional encryption quantum-resistant. Most security experts recommend a hybrid approach: deploy both QKD and PQC where possible to provide defense-in-depth against all known quantum threats.

Quantum computer

A quantum computer is a machine that performs computations using quantum mechanical phenomena such as superposition and entanglement. Unlike a classical computer, which processes bits that are always 0 or 1, a quantum computer manipulates qubits that can be in superposition. For certain specific problems, quantum algorithms run exponentially faster than any known classical algorithm. For most everyday computing tasks (browsing the web, running spreadsheets, sending email), a quantum computer offers no advantage. The cryptographic threat comes from two algorithms: Shor's algorithm for breaking asymmetric encryption, and Grover's algorithm for weakening symmetric encryption and hash functions. A quantum computer powerful and reliable enough to threaten current encryption is called a cryptographically relevant quantum computer (CRQC). Building a CRQC requires millions of low-error qubits, a milestone that has not yet been reached.

Quantum threat

The quantum threat is the risk that sufficiently powerful quantum computers will eventually break the asymmetric cryptographic algorithms (such as RSA and elliptic-curve cryptography) that protect most of today's digital communications. The threat has two time horizons: the future threat (when a cryptographically relevant quantum computer exists and can actively break encryption in real time) and the present threat from harvest-now-decrypt-later attacks (where recorded ciphertext can be decrypted retroactively once such a computer exists). The quantum threat motivates both the deployment of QKD for new key distribution and the migration of existing systems to post-quantum cryptographic algorithms.

Replay attack

A replay attack occurs when an attacker records a legitimate communication and later retransmits it to deceive one of the communicating parties. For example, an attacker who records a valid authentication message might replay it later to impersonate the original sender. In QKD, timestamps and session identifiers embedded in the classical channel messages are used to prevent replay attacks. Each key exchange session has a unique identifier; messages from a previous session cannot be replayed into a new one. Replay attacks are relevant to QKD not because they violate quantum physics, but because they could undermine the classical post-processing steps that turn raw quantum measurements into a usable key.

Shor's algorithm

Shor's algorithm is a quantum algorithm, published by mathematician Peter Shor in 1994, that can efficiently find the prime factors of a large number. This matters enormously for cryptography because the security of RSA (the most widely deployed asymmetric encryption scheme) rests on the assumption that factoring a large number into its prime components is computationally infeasible. On a classical computer, factoring a 2048-bit number would take longer than the age of the universe. On a sufficiently large quantum computer running Shor's algorithm, the same task could take hours or days. Shor's algorithm also breaks elliptic-curve cryptography. It is the primary reason the cryptographic community is urgently developing and deploying post-quantum alternatives.

Side-channel attack

A side-channel attack exploits information leaked by the physical implementation of a cryptographic system rather than the cryptographic algorithm itself. Timing information, power consumption, electromagnetic emissions, or acoustic noise can all reveal details about secret keys or operations. In QKD, side-channel attacks often target the single-photon detector: for instance, an attacker might exploit variations in the detector's response time to infer which polarization was measured. Side-channel attacks are a subset of implementation attacks and are a significant focus of practical QKD security research.

Trojan-horse attack

A Trojan-horse attack on QKD involves an attacker injecting bright light into Alice's transmitter (via the quantum channel, working backward) to illuminate the optical components inside and measure their settings. By analyzing the reflected light that returns from Alice's device, Eve can determine which basis and polarization Alice chose for each photon without ever touching the photons Alice sends to Bob. Trojan-horse attacks exploit the fact that optical fibers transmit light in both directions: the protocol sends photons only from Alice to Bob, but the physics allows light to travel from Bob's end toward Alice. Countermeasures include optical isolators that block reverse-traveling light and monitors that detect unusual levels of incoming light.

14 Appendix B: BB84 Step by Step

This appendix is a slow-motion tour through one complete BB84 key exchange. You will follow ten photons from Alice's laser to Bob's detector, watch them survive or get discarded, and then run the whole thing again with an eavesdropper in the middle. If you have a pencil and a blank piece of paper, pull them out now. No physics background is required. Every concept here was introduced in the main chapters; this appendix slows everything down so you can see each gear turning.

Optional math callouts appear at the end for readers who want the numbers behind the steps. They are clearly labeled so you can skip them without missing the core ideas.

The Cast and the Setup

Every BB84 exchange has two honest participants and one potential troublemaker.

Alice is the sender. She has a device that generates individual photons one at a time, and she can set each photon to carry a 0 or a 1. She can also choose how to orient that information. Think of orientation as choosing which kind of filter to put in front of the photon before she sends it. BB84 uses two possible filter types, and Alice picks one at random for each photon.

Bob is the receiver. He has a detector at the end of the fiber-optic cable. Bob does not know which filter Alice used for any given photon. He guesses randomly, placing his own filter in front of his detector before the photon arrives. When his filter matches Alice's, he reads the bit perfectly. When it does not match, physics gives him a coin-flip result that tells him nothing useful.

The two bases are the two filter orientations Alice and Bob can each choose from. The rectilinear basis, written as +, encodes bit 1 as a photon pointing straight up (|↑⟩) and bit 0 as a photon pointing sideways (|→⟩). The diagonal basis, written as X, encodes bit 1 as a photon tilted 45 degrees to the right (|↗⟩) and bit 0 as a photon tilted 45 degrees to the left (|↖⟩).

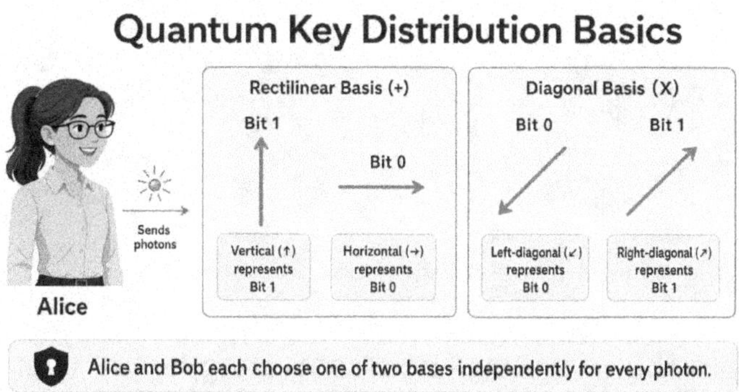

Diagram B.1: The two bases and four photon states

The key insight is simple. When Bob picks the same basis as Alice, his measurement yields the correct bit every time. When he picks the wrong basis, quantum mechanics forces his detector to produce a random result that is unrelated to what Alice sent. Bob cannot tell the difference from his end alone.

After the photons travel, Alice and Bob compare notes over a regular phone call or message that anyone can hear. They announce only which basis they used for each photon, never the bit values. They keep only the photons where they independently chose the same basis and throw away the rest. This step is called **sifting**, and it is how a stream of photon measurements turns into a shared key. Because anyone listening hears only the basis letters, not the bit values, the eavesdropper learns nothing about the secret.

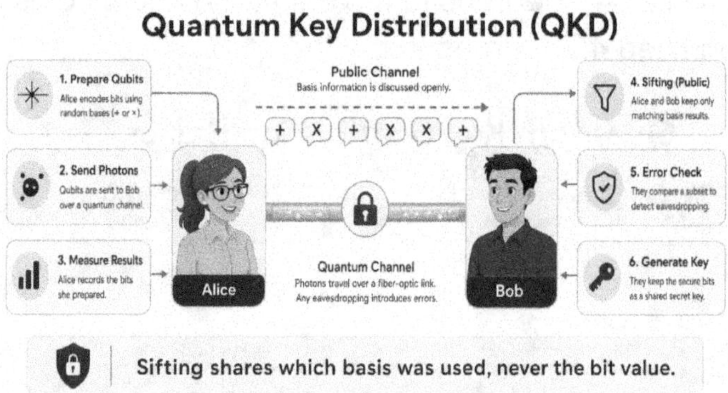

Diagram B.2: Alice and Bob comparing bases over a public channel

Step by Step with a Ten Photon Example

Below are ten photons. For each one, Alice picks a random bit and a random basis, then sends the matching photon state. Bob independently picks a random basis, measures, and records a bit. Work through these one at a time. When bases match, Bob is always correct. When they do not match, Bob's result is a random guess.

• **Photon 1:** Alice picks bit 0, basis +, sends |→⟩. Bob picks basis +. Bases match. Bob measures |→⟩, gets bit 0. KEEP.

• **Photon 2:** Alice picks bit 1, basis X, sends |↗⟩. Bob picks basis +. Bases do not match. Bob's + filter randomly collapses the photon; Bob gets bit 1 by chance. DISCARD. Bob cannot know whether his result matches Alice's, so neither party can trust it.

• **Photon 3:** Alice picks bit 0, basis X, sends |↖⟩. Bob picks basis X. Bases match. Bob measures |↖⟩, gets bit 0. KEEP.

• **Photon 4:** Alice picks bit 1, basis +, sends |↑⟩. Bob picks basis X. Bases do not match. Bob's X filter randomly collapses the photon; Bob gets bit 0 by chance. DISCARD.

• **Photon 5:** Alice picks bit 1, basis +, sends |↑⟩. Bob picks basis +. Bases match. Bob measures |↑⟩, gets bit 1. KEEP.

• **Photon 6:** Alice picks bit 0, basis X, sends |↖⟩. Bob picks basis X. Bases match. Bob measures |↖⟩, gets bit 0. KEEP.

- **Photon 7:** Alice picks bit 1, basis X, sends |↗⟩. Bob picks basis X. Bases match. Bob measures |↗⟩, gets bit 1. KEEP.
- **Photon 8:** Alice picks bit 0, basis +, sends |→⟩. Bob picks basis X. Bases do not match. Bob's X filter randomly collapses the photon; Bob gets bit 1 by chance. DISCARD.
- **Photon 9:** Alice picks bit 1, basis +, sends |↑⟩. Bob picks basis +. Bases match. Bob measures |↑⟩, gets bit 1. KEEP.
- **Photon 10:** Alice picks bit 0, basis X, sends |↖⟩. Bob picks basis +. Bases do not match. Bob's + filter randomly collapses the photon; Bob gets bit 0 with probability 1/2. DISCARD.

Read through those ten entries again. When the bases match, Bob reads exactly what Alice sent. When they do not match, Bob's result is a coin flip. He cannot tell whether it matches Alice's value from his result alone.

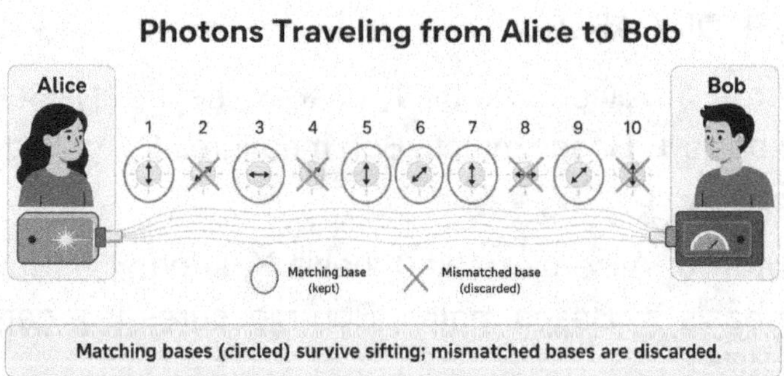

Diagram B.3: Ten photons in flight: bases matching and mismatching

The Sifting List

After the ten photons travel, Alice and Bob get on the phone and announce their basis choices one at a time. Anyone can listen. They only announce basis letters, not bit values.

- Photon 1: Alice used +, Bob used +. Match. KEEP.
- Photon 2: Alice used X, Bob used +. No match. DISCARD.
- Photon 3: Alice used X, Bob used X. Match. KEEP.
- Photon 4: Alice used +, Bob used X. No match. DISCARD.
- Photon 5: Alice used +, Bob used +. Match. KEEP.
- Photon 6: Alice used X, Bob used X. Match. KEEP.
- Photon 7: Alice used X, Bob used X. Match. KEEP.
- Photon 8: Alice used +, Bob used X. No match. DISCARD.
- Photon 9: Alice used +, Bob used +. Match. KEEP.
- Photon 10: Alice used X, Bob used +. No match. DISCARD.

Six photons survive: 1, 3, 5, 6, 7, and 9. Reading their bit values in order: 0, 0, 1, 0, 1, 1. That sequence is the **sifted key**. Alice has it in her notes. Bob has the same sequence in his notes because he measured each of those six photons with the correct basis. They now share the string 001011 without ever sending it as a readable message. Nobody who heard the public sifting call can reconstruct it from the basis letters alone.

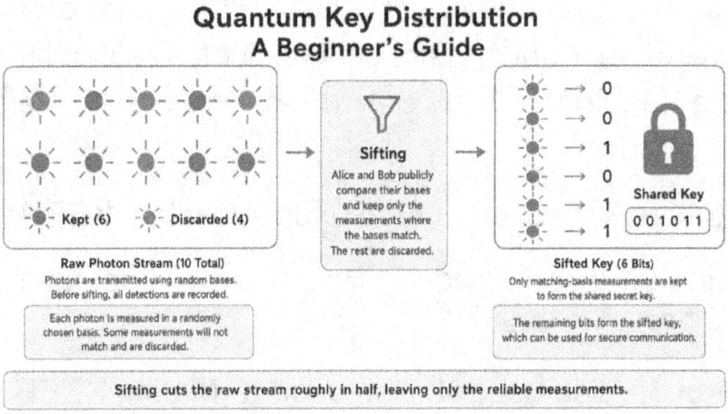

Diagram B.4: Before and after sifting: ten photons become six key bits

We started with ten bits and ended with six. In real deployments, engineers send thousands of photons to produce a useful key length. Ten photons is the smallest example that shows all the important features.

Optional math: Each person independently picks one of two bases, so the probability they both choose the same one is 1/2. Over a long run, roughly half the photons survive sifting. With ten photons, we expect five to survive on average. Seeing six in this example is normal random variation.

A Worked Eavesdropper Example

Now place an eavesdropper, **Eve**, in the middle of the channel. She intercepts each photon before it reaches

Bob, measures it with her own detector, and resends a replacement photon based on her measurement. She is trying to copy the key without being detected.

Eve faces the same problem Bob does. She does not know which basis Alice used. When she guesses wrong, she disturbs the photon. Her re-sent replacement carries her (wrong) measurement result in her (wrong) basis. When Bob later measures that replacement with the correct basis, the state has already been altered. Bob now has a 50 percent chance of getting the wrong bit, even on photons where he and Alice used matching bases.

Walk through the same ten photons with Eve present.

• **Photon 1:** Alice sent |→⟩ (bit 0, basis +). Eve guesses X (wrong). Eve gets a random result, say bit 1, and re-sends |↗⟩. Bob uses basis + (matching Alice). Bob's + filter randomly collapses Eve's diagonal photon; Bob gets bit 1. Alice sent 0, Bob recorded 1. ERROR.

• **Photon 2:** Discarded during sifting. Eve's interference here does not affect the key.

• **Photon 3:** Alice sent |↖⟩ (bit 0, basis X). Eve guesses X (correct). Eve measures bit 0 and re-sends |↖⟩. Bob uses X (matching Alice). Bob gets bit 0. No error. Eve was lucky.

• **Photon 4:** Discarded during sifting.

• **Photon 5:** Alice sent |↑⟩ (bit 1, basis +). Eve guesses + (correct). Eve measures bit 1 and re-sends |↑⟩. Bob uses + (matching Alice). Bob gets bit 1. No error. Eve was lucky again.

• **Photon 6:** Alice sent |↖⟩ (bit 0, basis X). Eve guesses + (wrong). Eve's + filter collapses the photon randomly; Eve gets bit 1 and re-sends |↑⟩. Bob uses X (matching Alice). Bob's X filter randomly collapses Eve's vertical photon; Bob gets bit 1. Alice sent 0, Bob recorded 1. ERROR.

• **Photon 7:** Alice sent |↗⟩ (bit 1, basis X). Eve guesses X (correct). Eve measures bit 1 and re-sends |↗⟩. Bob uses X (matching Alice). Bob gets bit 1. No error.

• **Photon 8:** Discarded during sifting.

• **Photon 9:** Alice sent |↑⟩ (bit 1, basis +). Eve guesses X (wrong). Eve's X filter collapses the photon randomly; Eve gets bit 0 and re-sends |↖⟩. Bob uses + (matching Alice). Bob's + filter collapses Eve's diagonal photon randomly; Bob gets bit 0. Alice sent 1, Bob recorded 0. ERROR.

• **Photon 10:** Discarded during sifting.

On the six kept photons, three contain errors (photons 1, 6, and 9). That is an error rate of 50 percent in this small example, higher than the theoretical average of 25 percent because small samples tend to vary widely. The important point is that in a run with thousands of photons, Eve's unavoidable guessing produces errors at close to 25 percent. When Alice and Bob compare a random sample of their sifted bits over the public channel, that error spike stands out immediately.

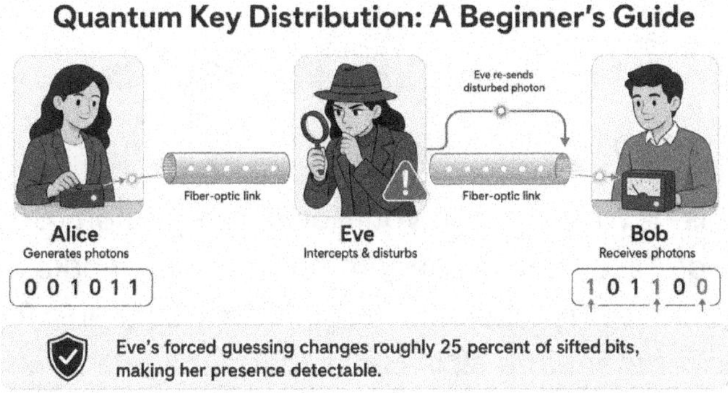

Diagram B.5: Eve intercepts and re-sends: how errors enter the sifted key

In a real BB84 session, Alice and Bob compare a subset of their sifted bits over the public call. If the error rate exceeds a threshold set above the normal noise floor of their channel, they abort the session, discard all the bits, and start over. Eve walked away with nothing usable, and Alice and Bob know the attempt happened.

This is the central security promise of BB84. Eve cannot measure a photon without disturbing it. She cannot make a perfect copy. She cannot read the key without leaving a fingerprint that Alice and Bob will find.

Optional math: Eve guesses the correct basis with probability 1/2. When correct, she introduces no error. When incorrect (probability 1/2), her re-sent photon gives Bob the wrong bit with probability 1/2. Expected error rate per kept photon: (1/2)(1/2) = 1/4, or 25 percent. Real channels add a few percent of noise, so implementations set the abort threshold above their channel's baseline.

A Blank Worksheet the Reader Can Photocopy and Run

Use this layout to run your own ten-photon BB84 exchange. A coin or a random number app works just as well for every choice. Fill in Alice's columns first, then cover them and fill in Bob's independently.

Alice's row entries (one row per photon, photons 1 through 10):

• Alice's bit choice: 0 or 1, chosen randomly

• Alice's basis choice: + or X, chosen randomly

• Photon state sent: |↑⟩ for bit 1 basis +; |→⟩ for bit 0 basis +; |↗⟩ for bit 1 basis X; |↖⟩ for bit 0 basis X

Bob's row entries (fill in without looking at Alice's columns):

• Bob's basis choice: + or X, chosen randomly

• Bob's recorded bit: copy Alice's bit if bases match; flip a coin if bases do not match

Sifting row entries (compare basis choices together):

• Bases match? Yes or No

• Keep or Discard

After sifting:

• List only the bit values for the kept photons, in order

• Alice's sifted key: write the sequence

- Bob's sifted key: write the sequence
- Do they match? They should match perfectly with no eavesdropper present

Optional second run with Eve:

- Add Eve's basis choice for each photon
- Mark where Eve's basis differs from Alice's
- Re-derive Bob's result using Eve's re-sent photon
- After sifting, count mismatches between Alice's and Bob's sifted keys

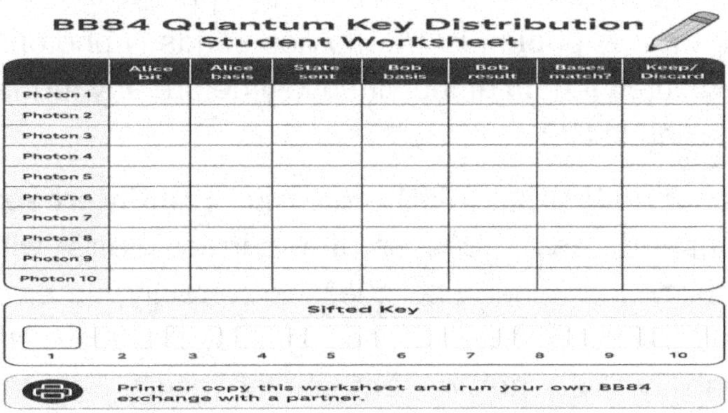

Diagram B.6: Blank worksheet layout for a ten-photon run

To make this vivid, try it with a partner. One person plays Alice, one plays Bob. A third person can secretly play Eve on the second run. Compare error rates between the two runs and watch the theory arrive on your notepad in pencil.

Optional Math Callouts

The following callouts collect the mathematical details flagged throughout the appendix. Each is self-contained. If you are comfortable with basic probability, read them in order. If you prefer to stay with the narrative, skip this section entirely.

Optional math: Probability of a basis match. Alice picks her basis from two choices (each with probability 1/2). Bob picks independently. The probability both choose + is (1/2)(1/2) = 1/4. The probability both choose X is also 1/4. Together, the probability of a match on any photon is 1/2. If Alice sends N photons, the expected length of the sifted key before any further processing is N/2.

Optional math: Expected error rate from intercept-and-resend. Eve guesses the correct basis with probability 1/2. When she guesses correctly, no error is introduced. When she guesses incorrectly (probability 1/2), her re-sent photon gives Bob the wrong bit with probability 1/2. So the probability of an error on any single kept photon is (1/2)(1/2) = 1/4, or 25 percent. Alice and Bob detect this by sampling even a few dozen sifted bits and checking whether the error rate is significantly above their channel's normal noise baseline.

Optional math: Partial eavesdropping and the detection threshold. Eve does not have to intercept every photon. If she intercepts a fraction f of the stream, the expected error rate in the sifted key drops to f/4. To stay below a 11 percent error threshold, Eve needs f/4 less than 0.11, meaning she can intercept at most roughly 44 percent of photons. But she gains useful information only from the roughly half of the intercepted photons for which she guessed correctly. Her usable information per sifted bit is therefore at most (0.44)(0.5) = 0.22, and privacy amplification removes even that fraction from the final key. There is no partial-intercept strategy that gives Eve a complete key while keeping Alice and Bob unaware.

Optional math: Why privacy amplification completes the job. After error correction, Alice and Bob agree on a key, but Eve may know some fraction of the bits. Privacy amplification applies a compression function to a reconciled key of length L, producing a shorter key of length L minus k, where k is an upper bound on how many bits Eve knows. Classical information theory guarantees that the compressed key carries essentially zero information about Eve's observations, even if she has unlimited computing power. BB84 provides the bound on k (derived from the observed error rate); privacy amplification then mathematically eliminates the leaked information. The result is a shared secret that no eavesdropper can reconstruct.

15 Appendix C: Free Simulators, Videos, and Resources

Learning about quantum key distribution (QKD) does not require expensive lab equipment or a university enrollment. A growing collection of free tools, videos, and hands-on activities lets you explore the core ideas at your own pace, on your own schedule, without spending a dollar. This appendix points you to the available resources and tells you what to look for when you search for them yourself.

The goal here is not to list every website or video that exists today. That would be a losing race against time, because links break, courses get updated, and new materials appear every year. Instead, this appendix describes what good resources look like, what search terms tend to surface them, and how to judge whether a resource is trustworthy before you spend an hour with it. Think of it as a compass, not a map.

You can treat these sections in any order. If you already feel comfortable with the BB84 protocol, skip straight to the home experiments or the course section. If you are beginning, start with the simulators and videos. There is no wrong path.

Interactive BB84 Simulators

The single most powerful thing you can do after reading about BB84 is watch it run. Several universities and science museums have built free, browser-based simulators that let you step through the protocol one photon at a time. No installation is required. You open the page, click a button, and the simulation begins.

When you search for a BB84 simulator, use terms like "BB84 interactive simulation," "quantum key distribution browser demo," or "BB84 step by step visualization." You will find options ranging from simple slide-show-style walkthroughs to full interactive environments where you control every variable.

The best simulators share a few features. Look for one that lets you choose how many photons to send, typically between 10 and 100. It should show Alice's bit choices and basis choices in a column or list next to Bob's basis choices so that you can see both sides of the conversation. After measurement, it should highlight which rounds used matching bases and which did not, and then show the resulting sifted key. The visualization of basis agreement is the most important part to study. Watching mismatched bases produce useless results makes the protocol click in a way that prose description alone cannot achieve.

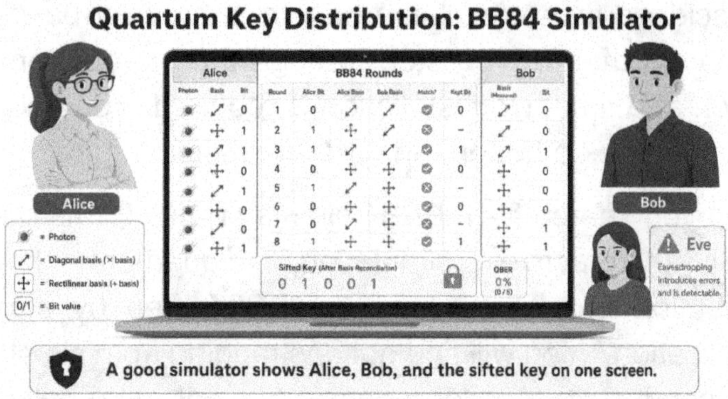

Diagram C.1: A browser-based BB84 simulator walkthrough

The most educational simulators also include an optional eavesdropper, often labeled Eve. When you toggle Eve on, the simulator introduces her measurement step between Alice and Bob. Watch how switching Eve on changes the error rate in the sifted key. A well-designed simulator will show you the percentage of bits that disagree when Eve is present and compare it against the clean run without her. This is the heart of QKD's security argument, made visible without a single equation.

A few simulators go further and let you adjust the quantum channel's natural error rate, sometimes called the quantum bit error rate (QBER). Raising the noise level and watching it blend with the Eve-induced errors is a useful exercise. It builds intuition for why the eavesdropping detection threshold exists and why QKD systems are calibrated carefully in real deployments.

If the simulator you find does not show Eve's effect, look for another one. The eavesdropper demonstration is the feature that matters most for understanding why the protocol is secure.

Reputable Explainer Videos and Short Courses

Video is often the fastest way to grasp a new concept, especially when it involves something you cannot see with your eyes, like a photon's polarization state. A well-made video combines animation, narration, and timed pacing in a way that reading alone cannot replicate.

Search for QKD explainer videos using terms like "quantum key distribution explained," "BB84 protocol animation," "quantum cryptography beginner," or "how quantum keys work." Add "no math" or "for beginners" to filter toward gentler introductions. Most major science communication channels on video platforms have covered QKD at least once, and a handful have done it exceptionally well.

When evaluating a video, look for these qualities. Does the narrator slow down and use an analogy before introducing a technical term? Does the animation match the narration, or is the animation generic stock footage that could illustrate anything? Does the video distinguish between QKD (a key distribution method) and quantum computing (a different technology)? Many casual videos blur that distinction badly, which leads to lasting confusion.

Short-course platforms also offer beginner-friendly introductions. Look for courses labeled "introduction to quantum computing" or "quantum information basics." Even though these courses cover more ground than QKD alone, the early modules on quantum bits, superposition, and measurement are directly relevant. If the course lists no prerequisites and promises to cover quantum states from scratch, it is likely appropriate for you.

Pay attention to the publication date or the last update date. Quantum technology moves quickly. A course or video from more than five years ago may still be accurate on the physics fundamentals, which do not change. Still, it may be outdated in terms of deployment status, real-world implementations, and security standards. Use older materials for the physics, and supplement them with newer materials on the current state of the field.

Polarization Demonstrations You Can Do at Home

You do not need a quantum lab to feel the physics. Polarization, the property that QKD uses to encode information onto photons, is observable with everyday objects. The demonstrations below use materials you likely already own or can find for a few dollars.

The two-lens demonstration. Find two pairs of polarizing sunglasses, or buy two small squares of polarizing film from a hobby or photography supplier. Hold one lens up to a lamp or window and look through it. The light passes through normally. Now hold the second lens in front of the first, with both lenses parallel and facing the same direction. Light still passes through. Slowly rotate the second lens ninety degrees relative to the first. The image darkens and eventually goes nearly black. You have just blocked almost all the light by orienting the two polarization filters at right angles to each other.

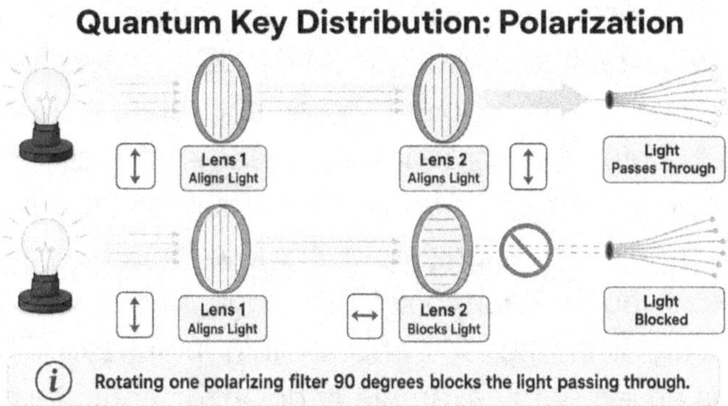

Diagram C.2: The two-lens polarization demonstration

This experiment makes one of the most important concepts in QKD tangible. When Alice and Bob choose the same measurement basis (both vertical, or both diagonal), the photon's information passes through correctly. When they choose incompatible bases, the result is essentially random, like looking through two perpendicular lenses and getting darkness. The sifting step in BB84 is the process of identifying the rounds in which Alice and Bob had their lenses aligned.

The diagonal basis extension. Once you have the two-lens setup, try this addition. Place the second lens at forty-five degrees instead of ninety degrees. Notice that some light passes through, but the image is dimmer than when the lenses were aligned and brighter than when they were perpendicular. This intermediate state is a real-world illustration of what happens when a photon prepared in one basis is measured in a different but non-perpendicular basis. The result is genuinely probabilistic, not simply blocked.

The LCD screen experiment. If you have a laptop or tablet with a liquid crystal display (LCD) screen, hold one polarizing lens up to the screen and rotate it slowly. At one angle, the screen looks nearly normal. At a perpendicular angle, it goes almost completely dark. LCD screens use polarizing layers to control how light passes through each pixel, so your single lens is interacting with the screen's built-in polarizer. This effect is an everyday example of polarized light at work in technology you use every hour.

The sky experiment. On a clear blue day, hold one polarizing lens up to a patch of blue sky (not directly toward the sun). Rotate the lens slowly. The sky will shift from lighter to noticeably darker blue as you rotate through ninety degrees. Sunlight scattered by the atmosphere is partially polarized, so your lens reveals that the sky itself carries information about the light's direction of travel. This effect would look like noise to an eavesdropper trying to intercept random photons without knowing the correct measurement angle.

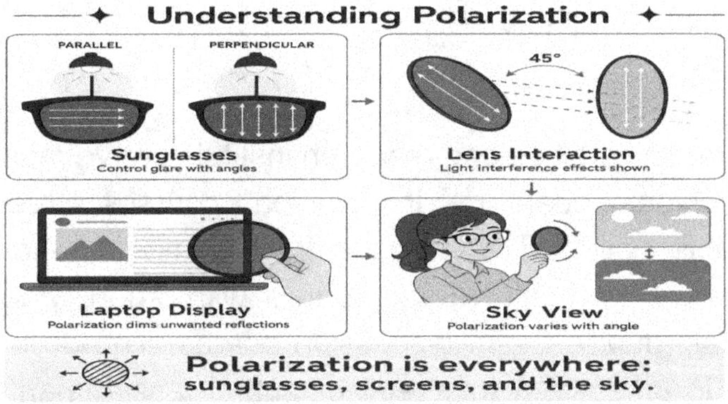

Diagram C.3: Four at-home polarization experiments side by side

These experiments will take no more than 20 minutes in total. After completing them, return to any BB84 simulator or video. You will find that the concepts land differently now. The language of bases, alignments, and random results will feel grounded in something physical you have held in your hands rather than abstract terms on a page.

One caution: cheap sunglasses sometimes use tinted lenses rather than genuine polarizing filters. To test whether a lens is polarizing, hold it up to a glossy surface at an angle. A true polarizing lens will show a noticeable change in the glare from the surface as you rotate it. If there is no change, the lens is not polarizing, and the experiment will not work well. Look for lenses or film explicitly labeled "polarized."

Free University and Institutional Courses

Several universities and research institutes have made introductory quantum information courses freely available online. These range from gentle one-hour overviews to multi-week courses with short video lectures, quizzes, and discussion boards. You do not need to take them for credit or submit assignments. Most allow open enrollment without registration.

When searching, use terms like "free quantum information course," "open quantum computing course beginners," "quantum cryptography MOOC," or "quantum key distribution lecture series." Many large open-course platforms aggregate offerings from universities worldwide. Filtering by "beginner" or "no prerequisites" will effectively narrow the results.

What should a good beginner course look like? The first week or module should introduce quantum bits (qubits) and explain superposition using a coin analogy or a similar everyday image, rather than Dirac notation or wave equations. The course should spend time on measurement and explain why measuring a quantum state disturbs it, because this is the foundation of QKD's security. By the end of the course, you should be able to explain BB84 in plain sentences without having to memorize any formula.

Avoid courses whose syllabi jump immediately to quantum gates, quantum circuits, or quantum error correction without a clear introductory module. Those are intermediate topics that already assume comfort with quantum bits. Starting there will feel like entering a foreign film midway through.

Institutional resources are also worth exploring. Several national laboratories and standards bodies publish free written guides and explainer documents aimed at business leaders and policy makers rather than physicists. These documents are written in plain language and focus on security implications rather than equations. Searching for "quantum cryptography primer" from a government standards body or a national research institute often turns up well-vetted, authoritative material.

University quantum computing centers frequently post recordings of public lectures from their seminar series. A talk titled "Introduction to quantum key distribution" given as part of a colloquium series is often more accessible than an academic paper on the same topic because the speaker is addressing a mixed audience of researchers from different fields rather than specialists.

Reputable News and Explainer Sites

Staying current with quantum technology does not require reading academic journals. A handful of specialized science journalism outlets and institutional blogs cover QKD and quantum cryptography accurately without being impenetrable.

When looking for news and updates, search for "quantum cryptography news," "QKD deployment updates," or "post-quantum cryptography standards" combined with the current year. Adding the current year to your search will surface recent developments rather than older announcements.

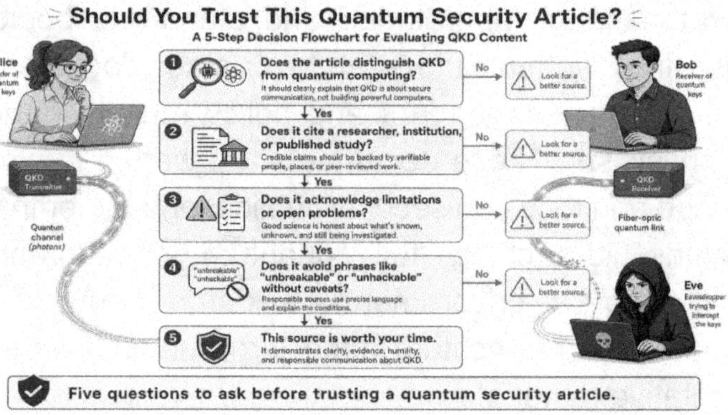

Diagram C.4: How to evaluate a quantum news source

Good science journalism on quantum topics does a few things consistently. It quotes researchers by their institutional affiliation rather than citing their social media presence. It distinguishes between what has been demonstrated in a lab, what has been deployed in a field trial, and what remains theoretical. It acknowledges that practical QKD systems face real engineering constraints around distance, cost, and key generation rate. Articles that describe QKD as simply "unbreakable" without any caveat are not accurate. The physics is sound, but real systems involve components, networks, and human decisions that introduce practical considerations.

Avoid aggregator sites that copy press releases verbatim without editorial review. These sites often amplify the most optimistic claims from a university press release without noting that the result was achieved under highly controlled lab conditions. The gap between a laboratory demonstration and a commercially deployable system can be very large. Good journalists explain that gap.

Institutional blogs from standards organizations are among the most reliable sources for policy-relevant information. These organizations have a mandate to provide balanced, peer-reviewed guidance and do not sell anything. Look for guidance documents on quantum-safe cryptography issued by recognized national or international standards bodies. These documents are updated regularly as the field develops and are freely downloadable.

Science museum websites and planetarium blogs occasionally publish long-form explainer articles on quantum technology. Because these organizations exist to educate the public, not to promote a product or a result, their content tends to be more balanced than corporate press releases. They also tend to date their articles clearly and note when an article has been updated.

How to Evaluate a Quantum Security Source

The quantum security space attracts a wide range of voices, from serious researchers with decades of peer-reviewed work to consultants whose primary credential is a conference appearance. Knowing how to separate reliable information from noise is a skill worth building early.

Start with the author's institutional affiliation. A researcher at a university physics department, a national laboratory, or a recognized cryptographic standards body has a professional reputation tied to accuracy. A blog post from an anonymous source or a vendor white paper written by a company selling quantum security products deserves more scrutiny.

Look at what the source acknowledges it does not know. Honest experts in quantum technology are comfortable with uncertainty. The field is moving rapidly, and anyone who writes as if all practical problems have been solved is either misinformed or overselling. Healthy skepticism about hype is a marker of expertise, not ignorance. If a source says "current systems can distribute keys over fiber at distances up to about a hundred kilometers under certain conditions," that precision signals credibility. If the source says "quantum networks will make all current encryption obsolete by next year," treat that claim carefully.

Check whether the source distinguishes between QKD and post-quantum cryptography (PQC). These are two different approaches to the same problem, and conflating them is a common error in lower-quality coverage. QKD uses physics to protect key distribution. PQC uses new mathematical structures that classical computers cannot easily break even with substantial processing power. A source that uses the terms interchangeably probably does not understand either one well enough to guide you.

Peer review is not a perfect filter, but it is a meaningful one. An approach or result that has survived peer review in a reputable physics or cryptography journal has been evaluated by independent experts who were not trying to make the result sound impressive. You do not need to read the journal article yourself to benefit from this filter. Simply noting that a result has been peer-reviewed and published in a named journal, as opposed to a preprint server or a conference talk alone, gives you a reasonable baseline of confidence.

Be aware of the distinction between the physics layer and the system layer. The BB84 protocol's security argument is mathematically rigorous. But a complete QKD system is more than the protocol. It includes photon sources, detectors, fiber or free-space channels, classical authenticated communication, software, and human operators. Claims that QKD is perfectly secure are claims about the protocol. Claims about practical security must address the whole system. A good source will be clear about which layer it is discussing.

Finally, notice how a source handles the eavesdropper problem. The claim that any eavesdropping on a QKD channel is detectable rests on specific assumptions, including that quantum mechanics is correct as currently understood, that the hardware behaves as specified, and that the authenticated classical channel used for basis comparison is genuinely secure.

Sources that acknowledge these assumptions are telling you the full picture. Sources that skip the assumptions and announce unconditional security are shortchanging you.

You now have a practical toolkit for exploring QKD beyond this book. The simulators, videos, experiments, courses, and evaluation skills in this appendix are all in service of the same goal: moving quantum key distribution from an abstract concept into something you have seen run, felt in your hands, and learned to evaluate with a critical eye. That combination of experience and judgment is worth more than any single source, including this one.

Key Concepts Checklist

• I can describe what to look for in a good BB84 browser simulator, including Alice and Bob's basis columns and the eavesdropper toggle.

• I understand why watching a simulator with Eve turned on is the most efficient way to see QKD's security argument in action.

• I can perform the two-lens polarization demonstration at home and explain what it shows about measurement bases.

• I know what distinguishes reliable quantum security journalism from vendor-driven or poorly sourced coverage.

• I can explain the difference between QKD and post-quantum cryptography (PQC) in plain terms.

• I understand why a beginner course should introduce qubits and superposition before discussing quantum gates or circuits.

Questions to Explore

1. After running a BB84 simulator both with and without Eve, what happens to the error rate when Eve measures every photon? What does that tell you about how many photons she would need to measure to stay undetected?

2. The two-lens polarization demonstration shows a gradual dimming as you rotate from zero to ninety degrees. What do you think happens if you rotate a third lens placed behind the first two at forty-five degrees? Try it and reason through what the result means.

3. When evaluating a news article about a new QKD record (distance, key rate, or speed), what three questions would you ask before accepting the claim as a meaningful advance?

4. Several national standards bodies have published guidance on transitioning from classical to quantum-safe cryptography. If you were advising a manager on which guidance to follow, how would you explain the difference between guidance from a standards body and guidance from a product vendor?

Try This / Reflect

Search for "BB84 simulator" on any major search engine and open the first two or three results. Run each one with ten photons and without Eve. Then run each one again with Eve present. Write down in plain words what you observe. Do the two simulators show the same error rate when Eve is active? If they differ, what might explain the difference?

After completing the two-lens home experiment, try explaining it to someone who has not read this book. Use only the words "filter," "angle," "light," "passes through," and "blocked." No physics vocabulary allowed. If you can explain the effect clearly without physics terms, you have genuinely understood it.

Glossary Terms

Quantum bit error rate (QBER) - The percentage of bits in the sifted key where Alice's and Bob's values disagree. A high QBER suggests either a noisy channel or an eavesdropper.

Sifted key - The set of bits that remain after Alice and Bob discard all rounds where they used different measurement bases.

Polarizing filter - A material that only transmits light waves oriented in one specific direction. Two filters at right angles to each other block nearly all light.

Basis - The measurement framework Alice or Bob uses to interpret a photon's state. If their bases match, the result is reliable. If they do not match, the result is random.

Post-quantum cryptography (PQC) - A family of mathematical encryption methods designed to be resistant to attacks by quantum computers, without requiring quantum hardware to run.

Peer review - A process in which independent experts evaluate a research result before it is published, providing a baseline check on accuracy and methodology.

Key Takeaways

• Browser-based BB84 simulators are the fastest way to see the protocol in action and are freely available through university and museum sites.

• The most valuable simulator feature is the ability to toggle an eavesdropper on and off and watch the error rate change.

• Polarization is observable at home with two polarizing lenses, a laptop screen, or a clear sky, and these experiments make abstract QKD concepts physically tangible.

• Reliable quantum security information comes from researchers with institutional affiliations, peer-reviewed results, and sources that acknowledge limitations.

• QKD and post-quantum cryptography (PQC) are different solutions to the same problem; confusing them is a common sign of an unreliable source.

16 Appendix D: A Reading Path

This appendix is a guided journey. The book you just finished was designed to give you the concepts without requiring any prior knowledge of physics or mathematics. But curiosity does not always stop there. You may want to go deeper to understand the equations behind the principles, read the original research papers, or track what standards bodies and practitioners say today. This appendix lays out a five-step path that takes you from this book toward that deeper territory, one layer at a time.

Each step describes what kind of resource it is, what you will gain, and what it will ask of you. The path is not a race. You can pause at any step and still be well-informed.

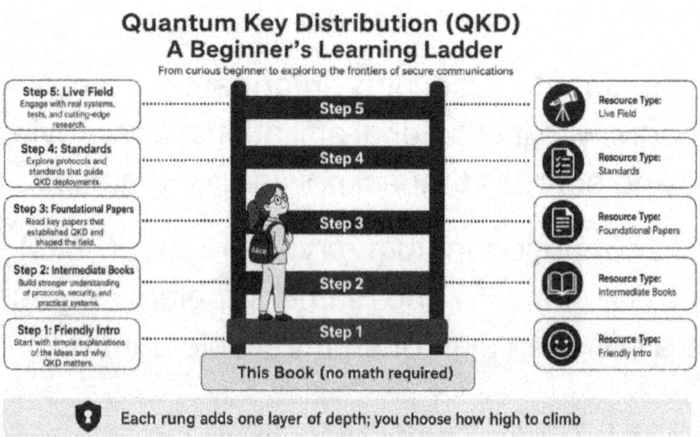

Diagram D.1: The five-step reading ladder

Step One: Other Friendly Introductions

The first step is about broadening your footing at the same level. After a book focused on quantum key distribution (QKD), you may find that surrounding ideas still feel incomplete. How does classical encryption work, and why does it fall short against quantum computers? What is quantum computing, and how is it different from QKD?

Friendly introductions to these adjacent topics appear in mainstream science publishing, technology journalism, and university science communication programs. You are looking for books or long-form articles with titles that use words like "explained," "a beginner's guide," or "for everyone." The best follow the same principle as this book: analogy first, mechanism second, equations never.

What you gain at this step is context. Reading at least two books that are not about QKD, one on quantum computing and one on classical cryptography, will give you a richer mental model of what researchers call the quantum-safe communications landscape. That phrase appears frequently in later steps, and it is far easier to absorb when you have already encountered each piece separately.

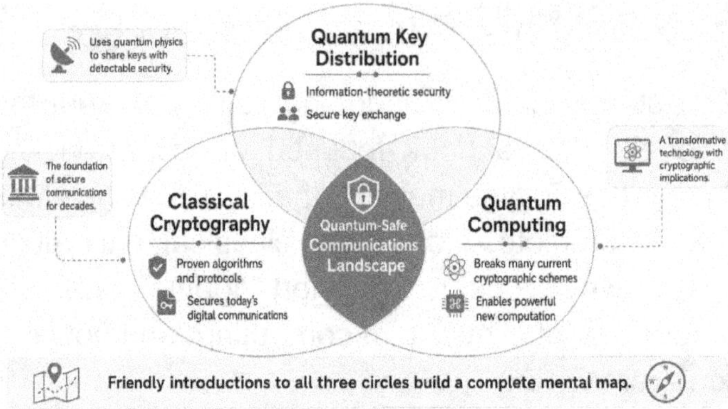

Diagram D.2: The three adjacent territories

Step Two: Intermediate Books and Articles

The second step is where the path begins to slope upward and mathematics first enters. Intermediate resources are written for readers who understand the concepts and want to see them expressed more precisely. The mathematics arrives gradually, as simple probability statements, basic algebra, or small equations placed alongside verbal explanations.

What you gain at this step is precision. You will move from understanding that photons carry information in their polarization states to being able to describe what that means in terms of angles and probability. You will move from understanding that eavesdropping disturbs the signal to being able to say something about the error rate that disturbance introduces.

The most important skill to develop here is reading around the mathematics when you are not ready for it.

Most intermediate texts are written so that a motivated non-specialist can follow the argument while skipping equations. Look specifically for resources that cover information theory alongside QKD, since concepts like entropy and error-correction codes are foundational to understanding how QKD works and its limits.

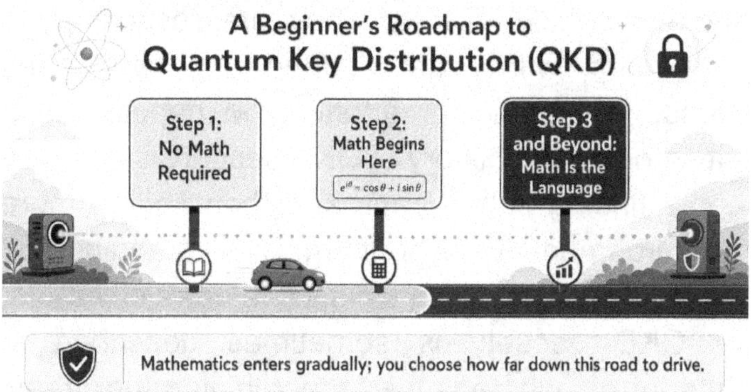

Diagram D.3: Where math enters the path

Step Three: Foundational Papers in Plain Context

The third step is where you encounter the original research. QKD has a short, traceable history. Its foundational ideas were proposed in a small number of papers written in the 1980s and 1990s. Because the field is young, those papers are still readable and still widely cited.

Before reading any foundational paper, seek out a secondary source that introduces it in plain language. These exist as blog posts, podcast episodes, recorded lectures, and book chapters. The secondary source explains why the paper was written, what problem it addresses, and why it matters today. With that context in hand, the paper itself becomes far more approachable.

What you gain at this step is a sense of how ideas develop in science. Foundational papers are not handed down fully formed. They are responses to earlier papers, with acknowledged gaps and open questions. Pay particular attention to papers about security proofs. The history of QKD security is a history of people finding subtle gaps and then finding more rigorous arguments to close them. Understanding that history will help you evaluate claims in news coverage, where QKD security is sometimes described in absolute terms, whereas the research literature treats it with much more nuance.

Step Four: Standards, Reports, and Agency Guidance

The fourth step moves from the laboratory to the policy room. QKD is not only a scientific idea. It is a technology that governments, standards organizations, and regulatory bodies are actively evaluating and in some cases deploying.

Committees write standards documents and agency reports. Their prose is dense, and their structures are formal. You look up the section you need, read it carefully, and note the version number and date, because these documents are updated regularly.

What you gain at this step is awareness of the official landscape. You will learn which algorithms have been selected or recommended for post-quantum cryptography, how national standards bodies assess the security claims of QKD vendors, and which open questions the standards community still considers unresolved.

You will also encounter the important distinction between QKD and post-quantum cryptography (PQC). These are two different approaches to the same problem. QKD uses the physics of quantum mechanics to generate and distribute keys. PQC uses mathematical problems believed to be hard even for quantum computers, but it runs on ordinary hardware. Standards bodies have made significant progress on PQC standards, while their guidance on QKD tends to be more cautious. Understanding why that difference exists is one of the most practically useful things you can learn at this step.

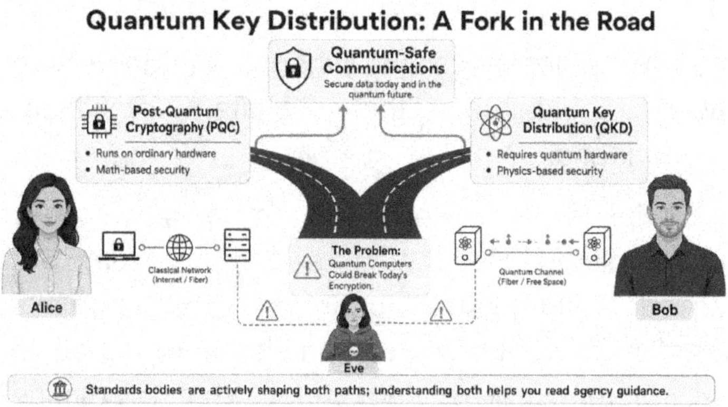

Diagram D.4: Two paths to quantum-safe communication

Step Five: Following the Field as It Evolves

The fifth step is not a destination. It is a practice. QKD is a rapidly developing field. The papers published this year will be cited and built upon next year. Standards released today may be revised in eighteen months as new experimental results arrive.

Following the field means establishing a small set of reliable ongoing sources and checking them regularly. The most useful are preprint repositories where researchers post new work before formal peer review, the websites of major standards bodies, the proceedings of leading venues in quantum cryptography, and the science coverage of trusted publications whose editors understand the difference between a laboratory demonstration and a commercially deployed system.

What you gain at this step is currency. You will know when a new security proof changes the bounds you learned in earlier steps and when a standards body releases guidance that shifts practitioner recommendations.

One discipline worth developing here is skepticism about headlines. When you read that a team has "achieved unbreakable quantum encryption" or "broken QKD," your first question should be: what was actually demonstrated, and under what conditions? The background you have built across the previous four steps will help you ask that question with confidence.

A Note on Pacing: Choosing Your Own Stopping Point

There is no obligation to reach Step Five. The reading path in this appendix is a map, not an itinerary.

If your goal was to be a more informed participant in conversations about cybersecurity and quantum technology, you have already reached it. This book, combined with one or two friendly introductions from Step One, is more than enough to follow news coverage intelligently and evaluate vendor claims with appropriate skepticism.

If your goal is to advise an organization on its security posture as quantum computers become more powerful, Steps Two and Four deserve the most attention. The intermediate technical background from Step Two will help you read the standards documents and agency reports of Step Four with real comprehension, not just surface familiarity.

If your goal is to work in the field as a researcher, engineer, or policy specialist, you will eventually want to work through all five steps and return to each repeatedly. The foundational papers of Step Three will become old friends. The standards documents for Step Four will serve as working references.

Whatever your goal, the most important thing is to start. The concepts in this book are the foundation.

Everything on this reading path builds on what you already know.

Key Concepts Checklist

• I understand that the reading path has five steps, each adding one layer of depth.

• I can identify what kind of resource belongs to each step.

• I understand that mathematics enters at Step Two and becomes the primary language by Step Three.

• I know that QKD and post-quantum cryptography (PQC) are two different approaches to the same problem, and that standards bodies treat them differently.

• I understand that I can choose my own stopping point based on my goals.

• I know that following the field as it evolves is a practice requiring ongoing attention.

Questions to Explore

1. What personal goal would drive how deep into this reading path you go? What would you want to be able to do or understand that you cannot do right now?

2. Why might it matter for a policy discussion whether an organization chooses QKD or post-quantum cryptography?

3. Foundational papers are described as responses to earlier papers, with acknowledged gaps. What does

that suggest about how you should read any single article about QKD?

4. Step Five emphasizes skepticism about headlines. Think of a recent headline about quantum technology. What questions would you ask to understand what was actually demonstrated?

Try This / Reflect

Find a recent article about QKD or quantum-safe communications from a mainstream technology or science publication. Read it and try to identify which step of the reading path it belongs to. Then look at the claims it makes. Are they presented with appropriate nuance, or do they oversimplify? Write two or three sentences describing what the article gets right and where a careful reader would want to know more.

Glossary Terms

Post-quantum cryptography (PQC) - A category of encryption methods designed to run on ordinary computers while remaining secure against attacks from quantum computers. Unlike QKD, PQC does not require quantum hardware on either end of the communication.

Preprint repository - An online archive where researchers post papers before or during formal peer review, making new results widely available quickly. Many QKD papers are accessible through such archives at no cost.

Review article - A paper written by researchers who survey and synthesize a large body of work in a subfield, designed to help newcomers understand the landscape without reading every primary source.

Standards body - An organization that develops and publishes formal technical specifications and recommendations. Standards bodies evaluate both QKD and PQC and issue guidance that practitioners rely on.

Security proof - A formal mathematical argument showing that a cryptographic protocol is secure under specified conditions and assumptions. The history of QKD includes a series of increasingly rigorous security proofs, each closing gaps identified in earlier versions.

Foundational paper - An original research article that introduced a concept, protocol, or technique the field subsequently built upon. In QKD, a small number of foundational papers from the 1980s and 1990s remain central references today.

Key Takeaways

• The reading path has five steps: friendly introductions, intermediate books and articles, foundational papers, standards and agency guidance, and following the evolving field.

• Mathematics enters gradually at Step Two and becomes central by Step Three; you choose how far along this spectrum to travel.

• Post-quantum cryptography and QKD are distinct approaches to quantum-safe communication, and understanding the difference is essential for reading standards documents intelligently.

• Foundational papers are accessible to motivated non-specialists when approached with secondary sources that provide context.

• Following the field as it evolves is a practice, not a destination: a small recurring investment of attention is more valuable than occasional deep dives.

17 Closing Thoughts

You made it. That is not a small thing.

When you opened this book, quantum key distribution (QKD) probably sounded like science fiction wrapped in acronyms. Photons carrying secrets. Physics as a lock. Eavesdroppers betrayed by the laws of the universe. It was a lot to take in, and you took it all in anyway. Give yourself a moment to appreciate that.

Now let us look back at the road you just walked.

The Journey, from the Beginning

You started with a simple question: how do two people share a secret when anyone might be listening? That question, it turns out, drives most of modern cryptography. Before QKD entered the picture, the best answer involved very large numbers and the assumption that factoring them was computationally hard. That assumption still holds today. But as you learned in Chapter 2, a powerful enough quantum computer could change that calculation, and some of the data being encrypted right now will still need to be secret decades from now. That is the "harvest now, decrypt later" threat, and it is the reason this conversation matters today, not someday.

Chapter 3 introduced the physics you needed: photons, their polarization states, and the core principle that measuring a quantum system disturbs it. That one idea, fragile as it sounds, turns out to be a security guarantee with no classical equivalent. You cannot copy an unknown quantum state without leaving evidence. You cannot eavesdrop without changing what you are eavesdropping on. The universe itself enforces these rules. No government, no court, no leaked password can override them.

Chapters 4 and 5 walked you through how those physics principles are applied to build a working key-sharing protocol. You followed Alice and Bob through the BB84 protocol step by step: choosing polarization bases, sending photons, comparing notes over a regular phone call, measuring the error rate, and reconciling their keys. You saw why a spy named Eve cannot hide. You also saw where human engineering can fall short even when the physics is perfect, because a photon detector that leaks timing information is a gap that has nothing to do with quantum mechanics and everything to do with careful engineering.

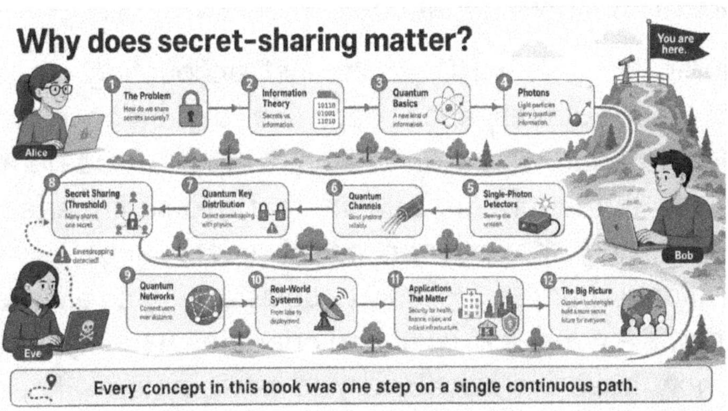

Diagram CL.1: The reader's journey through the book

Chapters 6 and 7 took QKD out of the lab and onto real fiber-optic cables, through metropolitan networks, and up to satellites. You learned about trusted nodes: the relay stations that make long-distance QKD possible today, but that also require you to trust the hardware at every stop. That is an honest limitation, and you now understand exactly why it exists and what it costs in terms of security guarantees. You also learned that satellites have changed the range game, allowing quantum keys to bridge distances that fiber alone cannot yet span reliably.

Chapters 8 and 9 put QKD to work in the sectors where it matters most right now: government and defense communications, financial institutions, healthcare data, and the critical infrastructure that keeps power grids and transportation systems running. These are not futuristic scenarios. Pilot networks in several countries are already protecting live traffic. The question is no longer whether QKD can be used in the real world. The question is how quickly costs come down, and standards mature so that deployment can scale.

Chapter 10 gave QKD an honest accounting of its limits: the authentication gap, the dependence on dedicated hardware and fiber, the trusted-node problem, the laboratory attacks that exploit imperfect devices, and the economics of cost, range, and operational burden. Security decisions are always trade-offs among risk, cost, and operational reality, and you now have the vocabulary to participate in those trade-offs intelligently rather than taking either the press release or the advisory at face value.

Chapter 11 introduced post-quantum cryptography (PQC) as QKD's partner rather than its competitor. PQC algorithms run on ordinary computers and are currently being standardized to replace the public-key algorithms that a quantum computer could threaten. QKD and PQC solve related problems from different angles, and the smartest organizations are planning for both. You now understand the difference, which puts you well ahead of most decision-makers who have heard the marketing claims without the underlying logic.

Chapter 12 looked forward. Quantum repeaters could eventually remove the trusted-node limitation. Entanglement-based protocols promise even stronger hardware-independence. A quantum internet, while still years or decades away, is a genuine research direction backed by serious national programs. You left that chapter knowing which future developments to watch for and which dramatic headlines to treat with a healthy wait-and-see attitude.

What You Can Do With What You Know

Knowledge without application is just stored weight. So here is what this understanding is actually worth to you right now.

You can read a news article about a "quantum-safe network" or an "unbreakable encryption system" and immediately ask the right questions. Is this QKD, PQC, or just a marketing label attached to something classical? Is it a lab demo or a deployed production network? What are the trusted nodes? What is the threat model it is actually solving? How does it handle the gap between photon-level physics and the messy reality of real cables, real hardware, and real human operators?

You can sit in a meeting where someone is proposing a quantum-security investment and contribute a grounded perspective. Not just enthusiasm, not just skepticism, but the specific questions that distinguish durable progress from hype. That is a rare skill, and you have it now.

You can evaluate vendors and advisors. When someone claims their product provides "unconditional security," you know what that phrase actually requires and what conditions it depends on in practice. You know about side channels. You know about the trusted-node problem. You know that "quantum" in a product name is not itself a security guarantee.

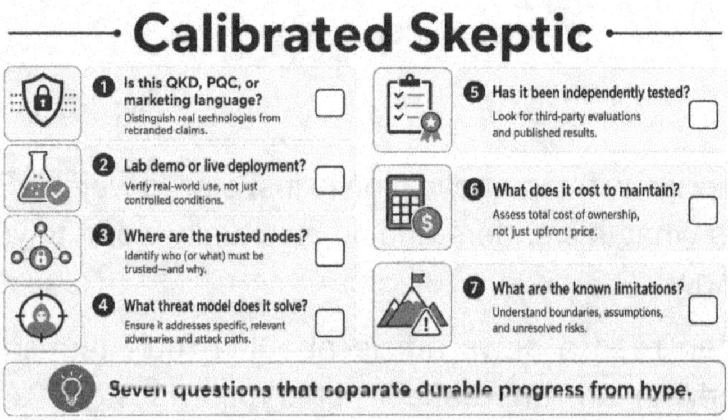

Diagram CL.2: Questions to ask any quantum-security claim

Most importantly, you can keep learning without starting from scratch. You have the conceptual scaffolding now. Every new article, conference talk, or product announcement about quantum security has a place to land in your mental model. That scaffolding is the real gift of this book, not any particular fact, but the structure that makes new facts make sense.

On Healthy Skepticism

One theme ran through every chapter, and it is worth naming directly before we part: honesty about what we know and what we do not.

QKD is real. The physics is real. The security guarantee at the photon level is genuinely unlike anything classical cryptography can offer. Deployed networks exist and are protecting live data in several countries right now.

QKD is also genuinely limited. Distance, cost, trusted nodes, the hardware gap, the coexistence problem with classical optical signals: these are real engineering constraints, not just temporary inconveniences that will disappear next year. Anyone who tells you otherwise is selling you something.

The quantum internet is a genuine long-term vision. Meaningful progress is happening. But "meaningful progress" and "available in five years" are very different sentences, and the headlines often blur that line.

Post-quantum cryptography is the near-term answer for most organizations. It runs on standard hardware, scales over the internet, and its security assumptions have been subjected to years of intense public scrutiny. For the vast majority of sensitive data, migrating to PQC-based systems is the most important quantum-security action you can take right now.

Holding all of these truths at once, without collapsing them into either "quantum security is everything" or "quantum security is hype," is calibrated skepticism. It is the most useful intellectual tool you can take away from this book.

Where to Go From Here

This book has three appendices, and they are not afterthoughts. They are the practical continuation of everything you just read.

Appendix B walks you through the BB84 protocol step by step with more technical detail than the main text allowed. If Chapter 4 left you wanting to see exactly how the sifting, error estimation, and privacy amplification steps work numerically, Appendix B is where that curiosity gets fed.

Appendix C is your gateway to continued learning. It lists free simulators you can run in a browser to watch QKD in action, video lectures by the researchers who built these systems, and resources to keep you current as the field evolves. The quantum-security landscape is moving. Appendix C is designed to move with it.

Appendix D is a reading path. It maps where to go next depending on your specific interest: deeper physics, practical cryptography, network architecture, policy and governance, or the long-term vision of the quantum internet. Whatever direction calls to you, Appendix D has a recommended route.

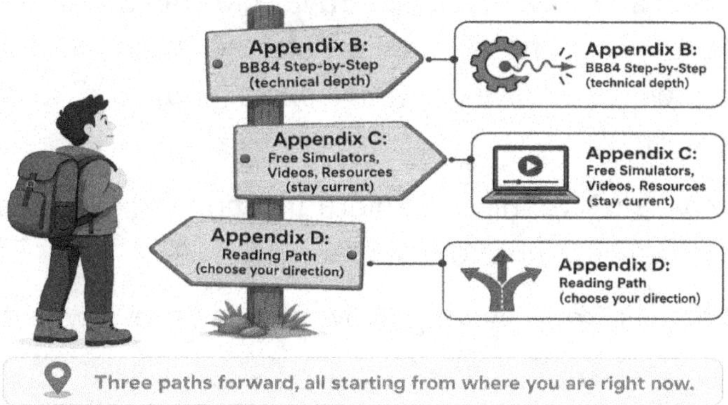

Diagram CL.3: Where to go next: a map to the appendices

A Final Word

You came to this book as a curious person who wanted to understand something real about the future of security. You are leaving it as someone who can hold a nuanced conversation, evaluate a claim, and teach the core ideas to someone else.

That last part matters. The ability to explain quantum key distribution in plain English to a colleague, a board member, or a curious friend is genuinely valuable. The gap between specialists and decision-makers in this field is wide, and people who can bridge it are rare. You can be one of them.

The physics of unbreakable keys is, at its heart, a beautiful idea: nature itself, not any algorithm, government, or corporation, enforces the secrecy of a shared secret. No amount of computational power, legal pressure, or insider betrayal can overcome the fact that measuring a quantum state changes it. That idea is worth knowing, worth explaining, and worth building on.

Thank you for reading. Explore the appendices. Stay curious. Ask the hard questions.

The field will keep moving. Now you can move with it.

18 References

A–D

Abellan, Carlos, et al. "Device-Independent Quantum Randomness Generation." *Nature*, 2023. (CH3, CH5)

Acín, Antonio, et al. "Quantum Networks: From Entanglement Distribution to the Quantum Internet." *Nature Reviews Physics*, 2024. (CH7, CH12)

Alléaume, Romain, et al. "Advances in Practical QKD Security Proofs." *npj Quantum Information*, 2024. (CH4, CH10)

Bacco, Davide, et al. "Metropolitan-Scale QKD with Integrated Photonics." *Nature Photonics*, 2023. (CH7)

Bai, Xue, et al. "Twin-Field QKD with Record Distance." *Physical Review Letters*, 2024. (CH4, CH7)

Banik, Subhadeep, et al. "Side-Channel Attacks in Quantum Cryptography." *IEEE Transactions on Quantum Engineering*, 2024. (CH10)

Bennett, Charles H., and Gilles Brassard. "BB84 at 40: Modern Perspectives." *Reviews of Modern Physics*, 2024. (CH4)

Brunner, Nicolas, et al. "Nonlocality and Device-Independent Security." *Nature Physics*, 2023. (CH5, CH10)

Cao, Yuan, et al. "Satellite-to-Ground QKD with Daylight Operation." *Nature Photonics*, 2024. (CH8)

Chen, Ji-Gang, et al. "Integrated Single-Photon Detectors for QKD." *Nature Electronics*, 2023. (CH6)

Ding, Dong, et al. "Quantum-Safe Financial Networks Using QKD." *Nature Communications*, 2024. (CH9)

E–H

Ecker, Sebastian, et al. "High-Dimensional QKD Over Fiber." *Nature Communications*, 2023. (CH4)

Elkouss, David, and Stefano Pirandola. "Quantum Repeaters: A 2024 Review." *Nature Reviews Physics*, 2024. (CH12)

Fan, Li, et al. "Chip-Scale Entangled Photon Sources for QKD." *Science Advances*, 2024. (CH6)

Feng, Zhi-Yuan, et al. "Security of QKD from Generalized Entropy Accumulation." *Nature Communications*, 2023. (CH4, CH10)

Gisin, Nicolas, et al. "Quantum Cryptography: State of the Art 2025." *Reviews of Modern Physics*, 2025. (CH1–12)

Google Gemini, All Images (CH1-CH18)

Grammarly, All Text Editing (CH1-CH18)

Gui, Ming, et al. "Quantum-Secured 6G Networks." *IEEE Communications Surveys & Tutorials*, 2024. (CH9, CH12)

Huang, Yu, et al. "Continuous-Variable QKD with Improved Noise Tolerance." *Physical Review Applied*, 2023. (CH4)

Hwang, Won-Young, et al. “Decoy-State QKD: 20-Year Update.” *Nature Photonics*, 2024. (CH4)

I–M

Islam, Nurul, et al. “Free-Space QKD in Urban Environments.” *Optica*, 2023. (CH7)

Jain, Nitin, et al. “Measurement-Device-Independent QKD: Practical Implementations.” *npj Quantum Information*, 2024. (CH5, CH10)

Kiktenko, Evgeniy, et al. “QKD for Critical Infrastructure Protection.” *Nature Communications*, 2024. (CH9)

Kumar, Rohit, et al. “Quantum-Resistant Key Exchange vs. QKD: A Comparative Study.” *ACM Computing Surveys*, 2024. (CH11)

Li, Wei, et al. “High-Rate Quantum Key Distribution Exceeding 110 Mb/s.” *Nature Photonics*, 2023. (CH6, CH7)

Liao, Sheng-Kai, et al. “Satellite QKD: Five-Year Operational Review.” *Nature*, 2024. (CH8)

Lin, Jiahui, et al. “Quantum-Secured Data Centers.” *IEEE Transactions on Cloud Computing*, 2024. (CH9)

Metger, Tony, and Renato Renner. “Security of QKD from Generalized Entropy Accumulation.” *Nature Communications*, 2023. (CH4, CH10)

Müller, Tobias, et al. “Quantum-Safe Network Architecture with Hybrid PQC + QKD.” *Nature Electronics*, 2024. (CH11)

N–S

Nauerth, Sebastian, et al. "Long-Distance Free-Space QKD with Moving Platforms." *Optics Express*, 2023. (CH7)

Pan, Jian-Wei, et al. "Entanglement-Based QKD at Metropolitan Scale." *Science*, 2024. (CH5, CH7)

Pironio, Stefano, et al. "Device-Independent QKD: Practical Security Bounds." *Physical Review Letters*, 2024. (CH5, CH10)

Pirandola, Stefano. "Quantum Communications Roadmap 2025." *Nature Photonics*, 2025. (CH12)

Qi, Bing, et al. "Advances in Continuous-Variable QKD Hardware." *Applied Physics Reviews*, 2024. (CH6)

Ren, Ji-Gang, et al. "Global-Scale Quantum Communication via Satellites." *Nature*, 2023. (CH8)

Sulimany, Kfir, et al. "High-Dimensional Coherent One-Way QKD." *npj Quantum Information*, 2025. (CH4)

Sun, Shuo, et al. "Quantum Dot Sources for Secure Communications." *Nature Materials*, 2023. (CH6)

T–Z

Tan, Si-Yu, et al. "Quantum Network Interoperability Standards." *IEEE Transactions on Quantum Engineering*, 2024. (CH12)

Tittel, Wolfgang, et al. "Quantum Repeaters: Experimental Progress." *Science Advances*, 2023. (CH12)